Lecture Notes in Physics

For information about Vols. 1–90, please contact your bookseller or Springer-Verlag.

Vol. 91: Computing Methods in Applied Sciences and Engineering, 1977, II. Proceedings, 1977. Edited by R. Glowinski and J. L. Lions. VI, 359 pages. 1979.

Vol. 92: Nuclear Interactions. Proceedings, 1978. Edited by B. A. Robson. XXIV, 507 pages. 1979.

Vol. 93: Stochastic Behavior in Classical and Quantum Hamiltonian Systems. Proceedings, 1977. Edited by G. Casati and J. Ford. VI, 375 pages. 1979.

Vol. 94: Group Theoretical Methods in Physics. Proceedings, 1978. Edited by W. Beiglböck, A. Böhm and E. Takasugi. XIII, 540 pages. 1979.

Vol. 95: Quasi One-Dimensional Conductors I. Proceedings, 1978. Edited by S. Barišić, A. Bjeliš, J. R. Cooper and B. Leontić. X, 371 pages. 1979.

Vol. 96: Quasi One-Dimensional Conductors II. Proceedings 1978. Edited by S. Barišić, A. Bjeliš, J. R. Cooper and B. Leontić. XII, 461 pages. 1979.

Vol. 97: Hughston, Twistors and Particles. VIII, 153 pages. 1979.

Vol. 98: Nonlinear Problems in Theoretical Physics. Proceedings, 1978. Edited by A. F. Rañada. X, 216 pages. 1979.

Vol. 99: M. Drieschner, Voraussage – Wahrscheinlichkeit – Objekt. XI, 308 Seiten. 1979.

Vol. 100: Einstein Symposion Berlin. Proceedings 1979. Edited by H. Nelkowski et al. VIII, 550 pages. 1979.

Vol. 101: A. Martin-Löf, Statistical Mechanics and the Foundations of Thermodynamics. V, 120 pages. 1979.

Vol. 102: H. Hora, Nonlinear Plasma Dynamics at Laser Irradiation. VIII, 242 pages. 1979.

Vol. 103: P. A. Martin, Modèles en Mécanique Statistique des Processus Irréversibles. IV, 134 pages. 1979.

Vol. 104: Dynamical Critical Phenomena and Related Topics. Proceedings, 1979. Edited by Ch. P. Enz. XII, 390 pages. 1979.

Vol. 105: Dynamics and Instability of Fluid Interfaces. Proceedings, 1978. Edited by T. S. Sørensen. V, 315 pages. 1979.

Vol. 106: Feynman Path Integrals, Proceedings, 1978. Edited by S. Albeverio et al. XI, 451 pages. 1979.

Vol. 107: J. Kijowski, W. M. Tulczyjew, A Symplectic Framework for Field Theories. IV, 257 pages. 1979.

Vol. 108: Nuclear Physics with Electromagnetic Interactions. Proceedings, 1979. Edited by H. Arenhövel and D. Drechsel. IX, 509 pages. 1979.

Vol. 109: Physics of the Expanding Universe. Proceedings, 1978. Edited by M. Demiański. V, 210 pages. 1979.

Vol. 110: D. A. Park, Classical Dynamics and Its Quantum Analogues. VIII, 339 pages. 1979.

Vol. 111: H.-J. Schmidt, Axiomatic Characterization of Physical Geometry. V, 163 pages. 1979.

Vol. 112: Imaging Processes and Coherence in Physics. Proceedings, 1979. Edited by M. Schlenker et al. XIX, 577 pages. 1980.

Vol. 113: Recent Advances in the Quantum Theory of Polymers. Proceedings 1979. Edited by J.-M. André et al. V, 306 pages. 1980.

Vol. 114: Stellar Turbulence. Proceedings, 1979. Edited by D. F. Gray and J. L. Linsky. IX, 308 pages. 1980.

Vol. 115: Modern Trends in the Theory of Condensed Matter. Proceedings, 1979. Edited by A. Pekalski and J. A. Przystawa. IX, 597 pages. 1980.

Vol. 116: Mathematical Problems in Theoretical Physics. Proceedings, 1979. Edited by K. Osterwalder. VIII, 412 pages. 1980.

Vol. 117: Deep-Inelastic and Fusion Reactions with Heavy Ions. Proceedings, 1979. Edited by W. von Oertzen. XIII, 394 pages. 1980.

Vol. 118: Quantum Chromodynamics. Proceedings, 1979. Edited by J. L. Alonso and R. Tarrach. IX, 424 pages. 1980.

Vol. 119: Nuclear Spectroscopy. Proceedings, 1979. Edited by G. F. Bertsch and D. Kurath. VII, 250 pages. 1980.

Vol. 120: Nonlinear Evolution Equations and Dynamical Systems. Proceedings, 1979. Edited by M. Boiti, F. Pempinelli and G. Soliani. VI, 368 pages. 1980.

Vol. 121: F. W. Wiegel, Fluid Flow Through Porous Macromolecular Systems. V, 102 pages. 1980.

Vol. 122: New Developments in Semiconductor Physics. Proceedings, 1979. Edited by F. Beleznay et al. V, 276 pages. 1980.

Vol. 123: D. H. Mayer, The Ruelle-Araki Transfer Operator in Classical Statistical Mechanics. VIII, 154 pages. 1980.

Vol. 124: Gravitational Radiation, Collapsed Objects and Exact Solutions. Proceedings, 1979. Edited by C. Edwards. VI, 487 pages. 1980.

Vol. 125: Nonradial and Nonlinear Stellar Pulsation. Proceedings, 1980. Edited by H. A. Hill and W. A. Dziembowski. VIII, 497 pages. 1980.

Vol. 126: Complex Analysis, Microlocal Calculus and Relativistic Quantum Theory. Proceedings, 1979. Edited by D. Iagolnitzer. VIII, 502 pages. 1980.

Vol. 127: E. Sanchez-Palencia, Non-Homogeneous Media and Vibration Theory. IX, 398 pages. 1980.

Vol. 128: Neutron Spin Echo. Proceedings, 1979. Edited by F. Mezei. VI, 253 pages. 1980.

Vol. 129: Geometrical and Topological Methods in Gauge Theories. Proceedings, 1979. Edited by J. Harnad and S. Shnider. VIII, 155 pages. 1980.

Vol. 130: Mathematical Methods and Applications of Scattering Theory. Proceedings, 1979. Edited by J. A. DeSanto, A. W. Sáenz and W. W. Zachary. XIII, 331 pages. 1980.

Vol. 131: H. C. Fogedby, Theoretical Aspects of Mainly Low Dimensional Magnetic Systems. XI, 163 pages. 1980.

Vol. 132: Systems Far from Equilibrium. Proceedings, 1980. Edited by L. Garrido. XV, 403 pages. 1980.

Vol. 133: Narrow Gap Semiconductors Physics and Applications. Proceedings, 1979. Edited by W. Zawadzki. X, 572 pages. 1980.

Vol. 134: $\gamma\gamma$ Collisions. Proceedings, 1980. Edited by G. Cochard and P. Kessler. XIII, 400 pages. 1980.

Vol. 135: Group Theoretical Methods in Physics. Proceedings, 1980. Edited by K. B. Wolf. XXVI, 629 pages. 1980.

Lecture Notes in Physics

174

Aida Kadić
Dominic G. B. Edelen

A Gauge Theory of Dislocations and Disclinations

Springer-Verlag
Berlin Heidelberg GmbH 1983

Authors

Dr. Aida Kadić
Gradjevinski Fakultet, Sarajevo, Yugoslavia

Dr. Dominic G. B. Edelen
Center for the Application of Mathematics
Lehigh University, Bethlehem, PA 18015, USA

This work was supported in part by the Marie Sklodowska-Curie Fund, U.S.-Polish Joint Board on Scientific and Technological Cooperation, N.S.F.-P.A.N. Grant No. J-F7F070-P

ISBN 978-3-540-11977-7 ISBN 978-3-540-39436-5 (eBook)
DOI 10.1007/978-3-540-39436-5

Originally published by Springer-Verlag Berlin Heidelberg New York in 1983

2153/3140-543210

TABLE OF CONTENTS

Abstract

Noting that the group $SO(3) \triangleright T(3)$ may be viewed as a 6-parameter gauge group that leaves the Lagrangian of elasticity theory invariant, the Yang-Mills universal gauge theory construction is used to erect a complete continuum theory of material bodies with dislocation and disclination fields. Breaking of the homogeneity of the action of $SO(3)$ is shown to give rise to disclinations and rotational dislocations while homogeneity breaking of $T(3)$ gives rise to translational dislocations. A rigorous justification for replacing displacement gradients by the components of the distortion tensor and Newtonian kinematic velocity by distortional velocity is obtained. A complete analysis of the theory is made and an expansion in terms of a group scaling parameter is obtained. It is shown that in the first order approximation classical elasticity theory is recovered, while the second and third orders of approximation model theories of dislocations and disclinations, respectively. Static solutions of the field equations in the linear engineering approximation reproduce the stress fields of edge and screw dislocations in the near field and exhibit exponential decay in the far field. Coupled systems of Klein-Gordon equations obtain in the dynamic case. The resulting dispersion relations provide direct means of determining the coupling constants by phonon experiments. Since the theory derives from a variational principle, it has a well defined momentum-energy tensor whose divergence vanishes for any and all solutions of the field equations. The additive structure of the Yang-Mills Lagrangian thus gives explicit expressions for the systems of forces and energy fluxes that act between the elastic, dislocation, and disclination modes of response. Particular cases give the Peach-Koehler forces on dislocations and the prediction that energy is lost by the elastic response whenever Drucker's postulate of plasticity theory is satisfied. Statements of the balance of linear and moment of momentum are shown to obtain as integrability conditions of the field equations.

Chapter 1

HISTORICAL REMARKS AND PHENOMENOLOGY OF DEFECTS

1.1 Historical Remarks

Dislocations were first discovered by Volterra [22], Weingarten [23], Somigliana [24] in the context of continuum mechanics. They developed the elastic properties of "macro"-dislocations in isotropic continua. It was only some 30 years later that the subject gained added importance through studies of discrete crystal lattices. Dislocations ceased to be mathematical curiosity when Orowan [25], Polanyi [26] and Taylor [27] predicted the existence of crystal imperfections that configurationally corresponded to dislocations. In 1934 they introduced the concept of an edge dislocation, a line-shaped crystal defect, into solid state physics. Five years later Burgers [28] discovered the second fundamental type of dislocations, the screw dislocations and developed what could be called the elastostatics of singular dislocations, both for isotropic and anisotropic media [29].

The subsequent development of dislocation theory can be divided roughly into three periods [30]. In the first of these periods, lasting until 1950, the theory was developed for simple dislocation configurations, usually straight dislocation lines, in infinite isotropic continua or in finite media with simple surface configurations. This period covers work done by the Italian School, and later by Orowan, Polanyi, Timpe, Taylor, Burgers, to mention just a few of the authors. In the second period, 1950-1965, extensive work in the theory of defects was done. Two major streams were developed; one mostly concerned with crystals and their properties in the context of dislocation theory, and the other trying to formulate continuum theories of dislocations and to bridge the gap between the discrete and continuum theories of defects. The latter were deemed important in studies of plastic deformations of materials. A number of theoretical embellishments were also introduced. It is not possible to list the names of all the authors who contributed to a better understanding of the various problems, we mention just very few of them, Kröner, Eshelby, Bilby, Bullough, Nye, Kondo, Nabarro, Mura.

The third period covers the years after 1965. It would be almost impossible to summarize all the work done in the last 15 years in a short paragraph. The interested reader is encouraged to see the review article by Kröner [10] for a detailed analysis of the present status of the theory of defects. In this period differential geometry became an integral part of the theory of dislocations and disclinations. Already in the early 1950's Kondo [31], followed independently by Bilby, Bullough and Smith [32], established the relation between dislocation theory and non-Riemannian geometry. Here the dislocation density plays the role of Cartan torsion [2]. Since the differential geometric representation of dislocation theory is the most natural and elegant one which is valid even for large deformations, it was widely accepted in the ensueing years. Also, group theoretical treatments of defect dynamics were suggested and some fundamental work was done. With all of these disciplines at hand, closely related to dislocations and disclinations, far-reaching analogies to the electromagnetic theory [33] the general theory of relativity [29], Yang-Mills theory [5] became of special interest.

Translational crystal defects, i.e. dislocations, were the primary object of study. Disclinations, i.e. rotational defects, did not attract significant attention, especially in studying crystals. It is much harder to detect disclinations experimentally. It seems that very large stresses are necessary in order to realize disclinations and it is expected that in such situations crystals would break. It is only when continuum theories are studied that rotational defects are sometimes taken into account. However, they are fully recognized in the studies of polymers, liquid crystals and amorphous bodies.

Gauge constructs in the theory of materials with defects are of quite recent vintage. The first occurrence in the literature is in a paper by A. A. Golebiewska-Lasota [33] in 1979. This was followed in rapid succession by three more papers [37, 38, 5] in which the gauge structure was considered in greater detail. Although the gauge groups considered were strictly Abelian, they were found to be much richer than the Abelian gauge group of classical electrodynamics; the general case having a 45-fold Abelian gauge group [5]. The obvious next question was that of the properties and structure of the corresponding gauge groups of the second kind. The suitability of Yang-Mills theory in answering such questions had already been remarked upon in [5]; the details and their implications are the subject of this work.

1.2 Phenomenology

A crystalline solid is often described as a perfectly regular array of atoms or molecules. Real crystals, however, are never so perfect. In point of fact, imperfections (i.e., defects) in the crystal lattice are responsible for many of the physical and chemical properties of solids.

A kind of crystal defect that has been studied intensively, particularly in metals and semiconductors is the dislocation. It is a line defect, well known to solid-state physicists, which plays an important role in plastic deformation of metals and many other phenomena besides. Associated with it is a translation. Namely, it results from a translation, or linear displacement, of one part of the crystal with respect to another part.

A defect associated with the rotational symmetry is called a disclination. In a disclinated solid one part of a structure is displaced relative to a neighboring part by a rotation rather than a translation.

Dislocations are studied more often than disclinations. They are found in conventional crystals. Disclinations are seldom observed in ordinary 3-dimensional crystals such as those of metals. They do appear in the arrays of oriented molecules called liquid crystals. Moreover, they are important structural elements in many ordered materials other than conventional crystals, such as the protein coats of viruses [34]. Disclinations can even be observed in the pattern of fingerprints, in the pelts of striped animals such as zebras and in basketwork. In recent years they have become important in studies of polymers and amorphous bodies.

Most often defects are studied in the context of crystalographic structures, so that the micro structure is essential. However, the continuum theory of defects, which is the macroscopic theory of the mechanical state of single and polycrystalline solids, is of value for the scientist since it provides the right mathematical formalism for handling all of the numerous mechanical problems that arise in attempts to understand the properties of solid materials from the crystalline constitution. The continuum theory should become of value also for the engineer who would be able to calculate material behavior from the given microscopic properties. Finally, the continuum theory is of great interest in itself since it is a very general and elegant field theory that can be a pattern for other field theories in modern physics.

Our primary concern is in the continuum field theory of defects. Hence, the phenomenology of dislocations and disclinations in crystal lattices will not be discussed here. It is assumed that the reader is familiar with the very basic ideas of defect dynamics. Nabarro [20] gave a detailed treatment of different types of defects. The reader is also strongly recommended to read the review article by Kröner [10].

Dislocations and disclinations can be viewed as topological defects. Their presence in bodies changes the topology, a simply connected region becomes multiply connected whenever there are defects. This, in turn, implies that the components of displacement are not single-valued functions whenever a defect line or surface is crossed. Hence, the ordinary theory of elasticity, which requires the displacements at any point of a body be single-valued functions of the coordinates of that point, ceases to be applicable in the theory of defects. And, yet, even after noting that displacements are multiple-valued, it is elasticity theory that was widely used in the past in obtaining results for problems of materials with defects.

Consider a doubly connected body represented by a sphere with a toroidal hole inside [20]. It has the property that a closed circuit described in the medium cannot shrink to a point while always remaining inside the medium. A single cut made outwards from the hole to the outer surface removes this property and the result is a topological sphere. The body may be dislocated by making a cut, displacing the two cut surfaces and possibly adding or removing thin layers of the medium, and finally joining the surfaces again in their disturbed positions. Choose two neighboring points close to the cut surface but on opposite sides. The reference coordinates of these points differ only infinitesimally, while their displacements differ significantly. In fact, on crossing the dislocation, i.e. the cut surface, the displacements have jump discontinuities. Relative displacements $\delta u^i(N)$ of neighboring points on either side of the cut at point N are given by the expressions [20]

$$\delta u^i(N) = b^i + d^i_j \, x^j(N) \, . \tag{1.2.1}$$

The relation (1.2.1) describes an incremental rigid body displacement of a point with respect to a neighboring point. The components d^i_j with

$$d^i_j = - d^j_i \, ,$$

can be considered as the infinitesimal generators of the 3-parameter rotation group SO(3). Similarly, the b^i's are viewed as the generators of the 3-parameter translation group T(3) . A semi direct product of these two groups gives a 6-parameter group $G = SO(3) \triangleright T(3)$ of all rigid body motions. Since the δu^i's are the relative displacements of particles on the top side of N, say, as compared with the particles on the bottom side of N, the δu^i's can be realized by allowing position dependent elements of the group G to act on going from the bottom side of N to the top side. The incremental displacement of the two sides of N is thus related by allowing the group G to act differently at different points, that is, to act inhomogeneously. The inhomogeneous action of this group at a generic point N is then given by (1.2.1). Nothing prevents us from extending this idea of the inhomogeneous action of G throughout the body. That is exactly what we do. We consider the inhomogeneous action of the group $SO(3) \triangleright T(3)$ globally, rather than locally.

At this point it seems to be useful to introduce Kröner's internal and external observers [10]. An internal observer deforms with a body, he cannot distinguish between compatible deformations. An external observer lives in the "external" space in which the body is embedded. He can see and do all the things that the inside observer sees and does and, in addition, he can distinguish between compatible deformations. Hence, the outside observer can compare distances from one compatible deformed state to the other which define strains and rotations. The role of the internal observer becomes important only when the body is filled with defects. Otherwise, although the outside observer sees a compatible distortion in the body without defects, the inside observer sees only vacuum which can be described by Euclidean geometry. It is the internal observer who perceives the essential physical events in the theory of defects. He is the one who detects the non-integrable distortions, for whom compatible deformations are not important. Indeed, the internal observer may be thought of as living in the space characterizing the inhomogeneous action of the group $SO(3) \triangleright T(3)$.

Once inhomogeneous action of the structure group $SO(3) \triangleright T(3)$ is allowed, great care must be exercised for we are playing with the elemental fabric of Newtonian mechanics. Indeed, each "particle" is endowed with three rotation and three translation degrees of freedom relative to neighboring particles so that only one particle at a time can be referenced by an inertial frame. Re-examinations of all of the basic ideas of Newtonian mechanics are thus called for if gross errors are to be avoided. A quest of such a fundamental nature is not an easy task without a sure guide through the myriad of possible alternatives. Fortunately, the calculus of variations and the Noether theorems provide a formalism of guaranteed self-consistency, for they relate invariance of the action functional under the action of the group with laws of balance that will be satisfied by all solutions of the field equations. Pursued in this manner, enforced invariance of the action functional under inhomogeneous action of $SO(3) \triangleright T(3)$ will guarantee satisfaction of the laws of balance of linear momentum and moment of momentum (Newton's laws). The only really new aspect is the enforcing instrumentality, namely the Yang-Mills gauge construct of high energy physics.

Chapter 2

PRELIMINARY CONSIDERATIONS

2.1 A Brief Review of the Exterior Calculus

The exterior calculus of E. Cartan provides a direct and natural basis for an analysis of the field equations of defect dynamics, for the basic field variables of the theory are differential forms. Since not all readers are equally familiar with this discipline, we give a brief review of the exterior calculus for forms defined on a 4-dimensional space E_4 . For a more detailed treatment the interested reader may consult references [1,2,3,4].

Let E_4 denote 4-dimensional Euclidean space referred to a given Cartesian coordinate cover $\{X^a, a = 1,2,3,4\}$. It is convenient to consider the four element index set $J = \{a\}$ as partitioned into a three element index set $I = \{A\}$ and a one element index set $\{4\}$; $\{a\} = \{A,4\}$. We shall also write $X^4 = T$, so that $\{X^a\} = \{X^A,T\}$.

The following conventions are adopted as to the values assumed by different indices. Whenever lower case Greek indices, $\alpha,\beta,\ldots,$ are used they take values from the set $I = \{1,2,3\}$. The same is true for indices denoted by capital letters, $A,B,\ldots,$ and by lower case Latin letters, $i,j,k,\ldots,$ starting with the letter i . Lower case Latin letters at the beginning of the alphabet, $a,b,\ldots,$ take their values from the index set $J = \{1,2,3,4\}$. The summation convention is adopted for repeated indices from any of the four kinds, the range of the implied summation being that of the kind of index involved. Thus $W^{\alpha}_{A} \gamma^{a}_{\alpha j} M^{A}_{a}$ stands for

$$\sum_{\alpha=1}^{3} \sum_{a=1}^{4} \sum_{A=1}^{3} W^{\alpha}_{A} \gamma^{a}_{\alpha j} M^{A}_{a}$$

and may thus be taken to define a quantity η_j with $j = 1,2,3$.

The collection of all smooth (C^{∞}) functions on E_4 is denoted by Λ^{o} . Such quantities are called scalars or forms of degree zero. The collection of all exterior differential forms of degree N , $N = 1,2,3,4$, on E_4 is denoted by $\Lambda^{N}(E_4)$. $\Lambda^{4}(E_4)$ is the 1-dimensional vector space of all 4-forms defined on E_4 with the natural basis $\{\pi\}$. π is also the volume element of E_4 , and is given by

(2.1.1) $$\pi = dX^1 \wedge dX^2 \wedge dX^3 \wedge dX^4 = \frac{1}{4!} e_{abef} \, dX^a \wedge dX^b \wedge dX^e \wedge dX^f ,$$

where e_{abef} are components of the Levi-Civita indicator tensor. A three-dimensional volume element of E_3 is denoted by $\mu = dX^1 \wedge dX^2 \wedge dX^3$, so that we obtain the elementary result

(2.1.2) $$\pi = \mu \wedge dT .$$

The symbol " $\wedge$ " is used to denote the exterior product.

The natural basis for $T(E_4)$, the tangent space of E_4, is given by the four linear operators $\{\partial_a, a = 1,2,3,4\}$, $\partial_a \equiv \frac{\partial}{\partial X^a}$. An element v of $T(E_4)$ can be written in the following way

$$v = v^a(X^b)\partial_a \quad .$$

The dual of the basis $\{\partial_1,\partial_2,\partial_3,\partial_4\}$ for $T(E_4)$ is the natural basis $\{dX^1,dX^2,dX^3,dX^4\}$ for the 4-dimensional vector space $\Lambda^1(E_4)$ of all 1-forms defined on E_4.

We need to introduce the top-down generated basis [3,5] for the vector spaces $\Lambda^3(E_4)$ and $\Lambda^2(E_4)$, in addition to the more customary natural basis. The top-down generated basis for the $\binom{4}{3}$-dimensional vector space $\Lambda^3(E_4)$ is given by

(2.1.3) $$\pi_a = \partial_a \rfloor \pi = \frac{1}{3!} e_{abef}\, dX^b \wedge dX^e \wedge dX^f \quad , \quad a\varepsilon J \; ,$$

where "$\rfloor$" stands for the operation of inner multiplication [3]. The properties of the elements of the set $\{\pi_a\}$ are

(2.1.4) $$d\pi_a = 0 \; , \quad dX^a \wedge \pi_b = \delta^a_b \pi \; ,$$

where $d = dX^a \wedge \partial_a$ denotes the 4-dimensional exterior derivative operator. A basis for the $\binom{4}{2}$-dimensional space $\Lambda^2(E_4)$ of all 2-forms defined on E_4 is given by

(2.1.5) $$\pi_{ab} = \partial_a \rfloor \pi_b = \partial_a \rfloor (\partial_b \rfloor \pi) \quad ,$$

for $a < b$. They have the properties

$$\pi_{ab} = -\pi_{ba}, \quad d\pi_{ab} = 0,$$

(2.1.6)

$$dX^c \wedge \pi_{ab} = \delta^c_a \, \pi_b - \delta^c_b \, \pi_a .$$

We deal with quantities defined in a 4-dimensional space, E_4 , where the first three coordinates $\{X^A\}$ are spatial and the 4 component is time. Hence, we need to relate them to the representations of the elements of $\Lambda(E_4)$ given above. The 4-dimensional exterior derivative operator can be written in the form

(2.1.7) $$d = dX^a \wedge \partial_a = dX^A \wedge \partial_A + dT \wedge \partial_4 = \bar{d} + dT \wedge \partial_4 ,$$

with $\bar{d} = dX^A \wedge \partial_A$ being the 3-dimensional exterior derivative operator. By observation, we conclude that

$$\partial_4 \rfloor \mu = 0, \quad \partial_4 \rfloor dT = 1, \quad \partial_A \rfloor dT = 0,$$

$$\partial_A \rfloor \mu = \mu_A ,$$

where $\{\mu_A\}$ is a top-down generated basis for the $\binom{3}{2}$-dimensional space $\Lambda^2(E_3)$. Therefore, from (2.1.2) and (2.1.3) together with (2.1.7) we obtain

$$\pi_a = (\partial_a \rfloor \mu) \wedge dT - \mu \wedge (\partial_a \rfloor dT) ,$$

that is

(2.1.8) $$\pi_a = \delta^A_a \, \mu_A \wedge dT - \delta^4_a \, \mu .$$

Now, any 3-form Ψ on E_4 is uniquely expressible in terms of the basis for $\Lambda^3(E_4)$ as follows

$$\Psi = \Psi^a \pi_a = \Psi^A \pi_A + \Psi^4 \pi_4 ,$$

which, according to (2.1.8) , can be written in the equivalent form

$$(2.1.9) \qquad \Psi = \Psi^A \mu_A \wedge dT - \Psi^4 \mu .$$

Simple calculation then gives

$$(2.1.10) \qquad \begin{aligned} d\Psi &= [\bar{d}(\Psi^A \mu_A) + \partial_4 \Psi^4 \mu] \wedge dT \\ &= (\partial_A \Psi^A + \partial_4 \Psi^4)\mu \wedge dT \end{aligned}$$

Let χ be a 2-form on E_4 . It can be expressed uniquely in terms of a 1-form $\rho = \rho_A dX^A$ and a 2-form $\eta = \eta^A \mu_A$ through the relation

$$(2.1.11) \qquad \chi = \rho \wedge dT + \eta = \rho_A dX^A \wedge dT + \eta^A \mu_A ,$$

and its exterior derivative is determined by [5]

$$(2.1.12) \qquad \begin{aligned} d\chi &= (\bar{d}\rho + \partial_4 \eta) \wedge dT + \bar{d}\eta \\ &= [\bar{d}(\rho_A dX^A) + (\partial_4 \eta^A)\mu_A] \wedge dT + \bar{d}(\eta^A \mu_A) . \end{aligned}$$

The operation d of exterior differentiation leads to the definition of two subspaces of the space $\Lambda(E_4)$ in a natural way. An element ν of $\Lambda(E_4)$ is said to be *closed* if and only if

$$d\nu = 0 .$$

An element ν of $\Lambda(E_4)$ is said to be *exact* if and only if there exists a $\sigma \varepsilon \Lambda(E_4)$ such that

$$\nu = d\sigma \ .$$

The collection of all exact (closed) elements of $\Lambda(E_4)$ forms a subspace $\mathcal{E}(E_4)$ $(\mathcal{C}(E_4))$ of $\Lambda(E_4)$ over $\mathbb{R}$, but not over $\Lambda^0(E_4)$, and [3]

$$\mathcal{E}(E_4) \subset \mathcal{C}(E_4) \ .$$

By the Poincaré lemma, if S is a region of E_4 that can be shrunk to a point in a smooth way (S is a starshaped region with respect to one of its points, called a center), then

$$\mathcal{C}(S) \subset \mathcal{E}(S)$$

Thus, if $d\nu = 0$ on S then there exists a σ on S such that $\nu = d\sigma$. It is essential that S be starshaped; if it is not, there need not exist a σ such that $d\nu = 0$ implies $\nu = d\sigma$.

The collection of all p-by-q matrices whose entries are exterior differential forms of degree k defined on the space E_4 is denoted by $\Lambda^k_{p,q}(E_4)$. Matrices of exterior forms are denoted by boldface letters. Hence, $\boldsymbol{\Omega} \varepsilon \Lambda^3_{3,1}(E_4)$ is a column matrix of 3-forms on E_4.

2.2 Antiexact Forms and Their Properties

The classes of closed and exact differential forms are very important in the study of the exterior calculus. There is, however, one further class of forms that is essential in the analysis of the equations of defect dynamics; namely, antiexact forms.

Let ω be an exterior form of degree k, which we may write as

(2.2.1) $$\omega = \omega_{a_1 \ldots a_k}(X^b)dX^{a_1} \wedge dX^{a_2} \wedge \ldots \wedge dX^{a_k} .$$

We define a linear *homotopy operator* H on any starshaped region S of E_4 with center (X_o^a) relative to a given coordinate cover by

(2.2.2) $$H\omega = \int_0^1 \lambda^{k-1} \mathcal{X} \lrcorner \tilde{\omega}(\lambda) d\lambda ,$$

where

(2.2.3) $$\mathcal{X} = (X^a - X_o^a)\partial_a ,$$

(2.2.4) $$\tilde{\omega}(\lambda) = \omega_{a_1 \ldots a_k}(X_o^b + \lambda(X^b - X_o^b))\ dX^{a_1} \wedge \ldots \wedge dX^{a_k} .$$

Although a standard tool in algebraic topology, the linear homotopy operator is usually encountered in the exterior calculus only in the proof of the Poincaré Lemma. The properties of H are as follows [3]:

(2.2.5)

(a) $H: \Lambda^k(S) \to \Lambda^{k-1}(S)$, $k \geq 1$, $H\Lambda^o(S) = 0$,

(b) $dH + Hd =$ identity for $k \geq 1$

$(Hdf)(X^a) = f(X^a) - f(X^a_o)$ for $k = 0$,

(c) $(HH\omega)(X^a) = 0$, $(H\omega)(X^a_o) = 0$,

(d) $HdH = H$, $dHd = d$,

(e) $X \lrcorner H = 0$, $HX\lrcorner = 0$.

According to (2.2.5) (b), any $\omega \varepsilon \Lambda^k(S)$ satisfies

(2.2.6) $\omega = dH\omega + Hd\omega$.

The element of $E(S)$ defined by

(2.2.7) $\omega_e = dH\omega$

is the *exact part* of the form ω . Elements of $\Lambda^o(S)$ have no exact part.

Let

(2.2.8) $\omega_a = Hd\omega = \omega - \omega_e$.

By (2.2.5) (c) $H\omega_a = 0$, so that ω_a belongs to the kernel of the linear operator H . The element ω_a of $\Lambda^k(S)$ is the *antiexact* part of ω . The collection of all antiexact elements of $\Lambda^k(S)$ is denoted by $A^k(S)$. Since elements of $\Lambda^o(S)$ have no exact part, we make the identification $\Lambda^o(S) = A^o(S)$ (see (2.2.5 b)). The collection of all antiexact forms on S is denoted by $A(S)$. The important thing about $A(S)$ is that it is a *submodule* of $\Lambda(S)$, in sharp contrast with the subspaces $C(S)$ and $E(S)$. Thus, in particular, $A(S)$ is closed under the operation of exterior multiplication. Hence, each $A^k(S)$ is a vector space over $\Lambda^0(k)$,

$$A^k(S) \wedge A^m(S) \subset A^{k+m}(S) ,$$

and exterior products of elements of ker H belong to ker H .

There is a theorem [3] that says that

$$(2.2.9) \qquad \Lambda^k(S) = d(A^{k-1}(S)) \oplus A^k(S) , \quad k \geq 1 ,$$

or equivalently

$$\Lambda^k(S) = E^k(S) \oplus A^k(S) ,$$

and that this direct sum decomposition is unique for a given center of S . If $\omega \in \Lambda^k(S)$ for $k \geq 1$, then $\omega \in \ker H$ and $(2.2.5)_b$ gives

$$(2.2.10) \qquad \omega = Hd\omega \qquad \forall\ \omega \in A^k(S) , \quad k \geq 1 .$$

Thus, *the linear homotopy operator is the inverse of the exterior derivative operator on the submodule* $A(S)$.

The results given above hold for any starshaped region S of any differentiable manifold. It is important to note, however, that the underlying manifold used throughout this work is the 4-dimensional Euclidean space E_4 which is starshaped with respect to any of its points as center. Thus, if the underlying domain is all of E_4, there are many choices for the center and hence many choices for the linear homotopy operator. Since the linear homotopy operator is central to the theory presented here, it is important to realize that the theory is covariant with respect to the choice of the center. Let $\Phi : E_4 \longrightarrow {}'E$ be a regular diffeomorphism between replicas of E_4, then Φ induces the pull back map $\Phi^* : \Lambda({}'E_4) \longrightarrow \Lambda(E_4)$. If H is defined on E_4, then Φ induces a linear homotopy operator $'H$ on $'E_4$ by [3]

$$(2.2.11) \quad \Phi^* \, {}'H = H\Phi^* ;$$

that is $'H = \Phi^{-1*} H \Phi^*$. If we now allow Φ to be the "translation" $'X^a = X^a + k^a$, then $'H$ has the center $(X_o^a + k^a)$. An identification of $'E_4$ with E_4 then serves to induce the new linear homotopy $'H$ on E_4 with the new center $(X_o^a + k^a)$.

2.3 Yang-Mills Minimal Coupling Theory

The theory of defect dynamics developed in this work is based upon an extension of the now classic Yang-Mills minimal coupling theory for semi-simple gauge groups. It is therefore useful to give a brief review of the Yang-Mills theory at this point, as an assist to the general reader. Those interested in the details of the Yang-Mills theory are referred to the review articles [6, 7, 8, 9].

Consider a Lagrangian $L_o(\psi, \partial\psi)$ that describes certain known systems of fields ψ. For purposes of discussion, we assume that the state variables may be organized as the components of a column matrix ψ. Let G_o be a Lie group of matrices that acts on the state vector $\psi(X^a)$ from the left,

$$(2.3.1) \qquad {}'\psi(X) = A\psi(X) \; , \quad \partial_a A \equiv 0 \; , \quad A \in G_o \; .$$

Although the group G_o is usually assumed to be semi-simple, we purposefully delay making this assumption until the point where it becomes necessary. Since the action of an element A of the group G_o is homogeneous, namely it acts on ψ at each point X in exactly the same manner, it follows that

$$(2.3.2) \qquad \partial_a {}'\psi(X) = A\, \partial_a \psi(X) \; ;$$

i.e., the process of partial differentiation and the action of G_o commute. An essential aspect of the theory is that the group G_o is an internal symmetry group (gauge group) of the theory in the sense that the Lagrangian function L_o is invariant under the action of G_o :

$$(2.3.3) \qquad L_o('\underset{\sim}{\psi}, \partial_a{}'\underset{\sim}{\psi}) = L_o(\underset{\sim}{A}\underset{\sim}{\psi}, \underset{\sim}{A}\partial_a\underset{\sim}{\psi}) = L_o(\underset{\sim}{\psi}, \partial_a\underset{\sim}{\psi})$$

for all $\underset{\sim}{A} \varepsilon G_o$.

Suppose that the transformations of the group G_o are now allowed to be space-time dependent. We denote this new group by G . The equation (2.3.1) is then replaced by

$$(2.3.4) \qquad '\underset{\sim}{\psi}(X^b) = \underset{\sim}{A}(X^b)\ \underset{\sim}{\psi}(X^b) .$$

Due to the inhomogeneity of the action of the group G , we now have

$$(2.3.5) \qquad \partial_a{}'\underset{\sim}{\psi} = \partial_a(\underset{\sim}{A}\underset{\sim}{\psi}) = (\partial_a\underset{\sim}{A})\underset{\sim}{\psi} + \underset{\sim}{A}(\partial_a\underset{\sim}{\psi}) ,$$

instead of the simple commutation law (2.3.2). Thus $\underset{\sim}{A}$ no longer factors from the left and the invariance of the Lagrangian L_o is lost:

$$L_o('\underset{\sim}{\psi}, \partial_a{}'\underset{\sim}{\psi}) = L_o(\underset{\sim}{A}\underset{\sim}{\psi}, (\partial_a\underset{\sim}{A})\underset{\sim}{\psi}+\underset{\sim}{A}(\partial_a\underset{\sim}{\psi})) \neq L_o(\underset{\sim}{\psi}, \partial_a\underset{\sim}{\psi}) .$$

Notice that the action of the group G does not change the coordinate cover, so in particular $'\partial_a \equiv \partial_a$ (i.e., the group G is a gauge group).

It is clear that preservation of the invariance of the Lagrangian L_o under the inhomogeneous transformations (2.3.4) requires something other than just the simple replacement of $\partial_a \underset{\sim}{\psi}$ by $\partial_a {}'\underset{\sim}{\psi}$. We would like to find an operator D_a such that

$$(2.3.6) \qquad {}'D_a {}'\underset{\sim}{\psi}(\underset{\sim}{X}) = \underset{\sim}{A}(\underset{\sim}{X})\; D_a \underset{\sim}{\psi}(\underset{\sim}{X}) \qquad \forall \underset{\sim}{A} \in G \; .$$

Whenever the state vector is transformed inhomogeneously, let us define the operator D_a and its image $'D_a$ by

$$(2.3.7) \qquad D_a \underset{\sim}{\psi} = \partial_a \underset{\sim}{\psi} + \underset{\sim}{\Gamma}_a \underset{\sim}{\psi} \; ,$$

and

$$(2.3.8) \qquad {}'D_a {}'\underset{\sim}{\psi} = \partial_a {}'\underset{\sim}{\psi} + {}'\underset{\sim}{\Gamma}_a {}'\underset{\sim}{\psi} \; , \quad a \in \{1,2,3,4\} \; .$$

Here, $\{\underset{\sim}{\Gamma}_a\}$ is a set of new fields, the so called compensating or gauge fields, that transform under the action of the group G according to

$$(2.3.9) \qquad {}'\underset{\sim}{\Gamma}_a = \underset{\sim}{A}\underset{\sim}{\Gamma}_a \underset{\sim}{A}^{-1} - (\partial_a \underset{\sim}{A})\underset{\sim}{A}^{-1} \qquad \forall \underset{\sim}{A} \in G \; .$$

The operator D_a is called the covariant derivative and $\underset{\sim}{\Gamma}_a$ are the corresponding connection matrices. A simple calculation then shows that $\underset{\sim}{A}$ again factors on the left when acting on $D_a\underset{\sim}{\psi}$ and hence (2.3.6) holds. Thus the "minimal" replacement

$$\partial_a\underset{\sim}{\psi} \rightarrow D_a\underset{\sim}{\psi} \tag{2.3.10}$$

restores the invariance of the Lagrangian L_o, for we then have

$$L_o('\underset{\sim}{\psi}, 'D_a'\underset{\sim}{\psi}) = L_o(\underset{\sim}{A}\underset{\sim}{\psi}, \underset{\sim}{A}D_a\underset{\sim}{\psi}) = L_o(\underset{\sim}{\psi}, D_a\underset{\sim}{\psi}) \tag{2.3.11}$$

by (2.3.3).

The connection matrices $\underset{\sim}{\Gamma}_a$, $a\varepsilon\{1,2,3,4\}$, are not as arbitrary as might appear, for (2.3.9) shows that they take their values in the Lie algebra G associated with the Lie group G, whose generators are constant matrices $\underset{\sim}{\gamma}_\alpha$, $\alpha = 1,2,\ldots,r$. Therefore, each of the $\underset{\sim}{\Gamma}_a$'s can be expressed in terms of the basis for the algebra G by

$$\underset{\sim}{\Gamma}_a = W_a^\alpha(\underset{\sim}{X})\underset{\sim}{\gamma}_\alpha, \tag{2.3.12}$$

where the W_a^α's are the Yang-Mills potential functions associated with the inhomogeneous action of the gauge group G. The matrices $\underset{\sim}{\gamma}_\alpha$ are subject to the commutation relation

$$[\underset{\sim}{\gamma}_\alpha, \underset{\sim}{\gamma}_\beta] \equiv \underset{\sim}{\gamma}_\alpha\underset{\sim}{\gamma}_\beta - \underset{\sim}{\gamma}_\beta\underset{\sim}{\gamma}_\alpha = C_{\alpha\beta}^\varepsilon\, \underset{\sim}{\gamma}_\varepsilon, \tag{2.3.13}$$

where $C^{\varepsilon}_{\alpha\beta}$ are the structure constants of the Lie algebra G . They satisfy the relations

$$(2.3.14)\quad C^{\varepsilon}_{\alpha\beta} = - C^{\varepsilon}_{\beta\alpha} ,$$

and the Jacobi identities

$$(2.3.15)\quad C^{\delta}_{\alpha\beta}C^{\varepsilon}_{\delta\gamma} + C^{\delta}_{\beta\gamma}C^{\varepsilon}_{\delta\alpha} + C^{\delta}_{\gamma\alpha}C^{\varepsilon}_{\delta\beta} = 0 .$$

For semisimple groups, the Cartan-Killing metric

$$(2.3.16)\quad C_{\alpha\beta} = C^{\delta}_{\alpha\gamma}C^{\gamma}_{\beta\delta} = C_{\beta\alpha}$$

form the components of a nonsingular matrix.

The replacement (2.3.10), that is necessary in order to preserve the invariance of the Lagrangian under the inhomogeneous action of the group G , gives rise to the new fields $\{W^{\alpha}_{a}(X^{b})\}$ that couple to the original state fields $\{\psi^{i}(X^{b})\}$ as follows

$$(2.3.17)\quad D_{a}\underset{\sim}{\psi} = \partial_{a}\underset{\sim}{\psi} + W^{\alpha}_{a}\, \underset{\sim}{\gamma}_{\alpha}\underset{\sim}{\psi} .$$

Additional field equations are thus required for determination of the Yang-Mills fields $\{W^{\alpha}_{a}\}$.

We proceed by replacing the original Lagrangian, now $L_{o}(\underset{\sim}{\psi}, D_{a}\underset{\sim}{\psi})$, by the new Lagrangian

$$(2.3.18)\quad L = L_{o}(\psi^{i}, D_{a}\psi^{i}) + s_{1}L_{1}(W^{\alpha}_{a}, \partial_{b}W^{\alpha}_{a}) ,$$

where the new term L_1 depends only on the potential fields and their derivatives and is required to be invariant under the inhomogeneous action of the gauge group G. Here, s_1 is a coupling constant. Variation of the action functional with respect to the W^α_a's gives the new field equations for the determination of the Yang-Mills fields, while variation with respect to the $\underset{\sim}{\psi}$'s gives the field equations for the $\underset{\sim}{\psi}$'s that are minimally coupled to the W^α_a-fields.

The construction of the G- invariant Lagrangian L_1 is significantly simplified if we rewrite (2.3.12) as a matrix of 1-forms by

$$(2.3.19) \quad \underset{\sim}{\Gamma} = \underset{\sim}{\Gamma}_a dX^a \ , \quad W^\alpha = W^\alpha_a \, dX^a \ , \quad \underset{\sim}{\Gamma} = W^\alpha \underset{\sim}{\gamma}_\alpha \varepsilon \Lambda^1_{3,3}(E_4) \ ,$$

where the W^α's are now the Yang-Mills 1-form potentials. The curvature matrix of 2-forms, $\underset{\sim}{\Theta}\varepsilon\Lambda^2_{3,3}(E_4)$, associated with the connection 1-forms $\underset{\sim}{\Gamma}$ is given by [1,2]

$$(2.3.20) \quad \underset{\sim}{\Theta} = d\underset{\sim}{\Gamma} + \underset{\sim}{\Gamma} \wedge \underset{\sim}{\Gamma}$$

and (2.3.9) shows that it transforms under the action of G by

$$(2.3.21) \quad {}'\underset{\sim}{\Theta} = \underset{\sim}{A}\underset{\sim}{\Theta}\underset{\sim}{A}^{-1} \ ,$$

where we have used the fact that $d = dX^a \wedge \partial_a$. Furthermore, (2.3.13), (2.3.19) and (2.3.20) show that

$$(2.3.22) \quad \underset{\sim}{\Theta} = F^\alpha \underset{\sim}{\gamma}_\alpha \ , \quad F^\alpha = \frac{1}{2} F^\alpha_{ab} \, dX^a \wedge dX^b \ \varepsilon \Lambda^2(E_4) \ ,$$

with $\{F^\alpha_{ab}\}$ the components of the Yang-Mills field tensor. The explicit evaluation is

$$(2.3.23) \quad F^\alpha = dW^\alpha + \frac{1}{2} C^\alpha_{\beta\gamma} \, W^\beta \wedge W^\gamma \ .$$

Now, (2.3.21) shows that $\underset{\sim}{\Theta}$ transforms according to the adjoint representation of G and (2.3.22) and (2.3.23) show that it depends only on the W^α_a's and their partial derivatives, being linear in the latter. A direct analogy with electrodynamics and a restriction to Lagrangian functions that are gauge invariant and quadratic in the derivatives leads to a construction of L_1 :

$$(2.3.24) \quad L_1 = \frac{1}{2} C_{\alpha\beta} \, F^\alpha_{ab} \, g^{ac} \, g^{bd} \, F^\beta_{cd} \ .$$

Here, $C_{\alpha\beta}$ are the components of the Cartan-Killing form on the group G . It is at this point that one requires G to be semi-simple in order to guarantee that L_1 is not a degenerate Lagrangian; i.e., $\det(C_{\alpha\beta}) \neq 0$ so that $((C_{\alpha\beta}))$ has an inverse. In point of fact, the minimal replacement construct described above is valid for arbitrary matrix Lie groups if the nondegeneracy condition is relaxed. The components of the tensor g^{ab} are given by

$$(2.3.25) \quad g^{AB} = -\delta^{AB}, \quad g^{44} = \frac{1}{\zeta}, \quad g^{ab} = 0 \Leftrightarrow a \neq b .$$

In contrast with electrodynamics, there is no requirement that ζ be equal to the speed of light *in vacuo* unless L_1 be required to exhibit Lorentz invariance. It is also useful to note that (2.3.24), subject to the conditions (2.3.25), can be obtained directly by standard isotropy and homogeneity arguments of classical linear continuum theories together with the gauge invariance condition.

The covariant exterior differentiation operator D is defined by

$$(2.3.26) \quad D = dX^a \wedge D_a .$$

If $\underset{\sim}{\eta}$ is a matrix of k-forms for which the action of the group G gives $'\underset{\sim}{\eta} = \underset{\sim}{A}\underset{\sim}{\eta}$, $\underset{\sim}{A} \varepsilon G$, then the covariant exterior derivative

$$(2.3.27) \quad D\underset{\sim}{\eta} = d\underset{\sim}{\eta} + \underset{\sim}{\Gamma} \wedge \underset{\sim}{\eta}$$

is transformed by the action of G according to

$$(2.3.28) \quad 'D('\underset{\sim}{\eta}) = \underset{\sim}{A}\, D\underset{\sim}{\eta} .$$

If $\underset{\sim}{\rho}$ is a matrix of k-forms that transforms under G by $'\underset{\sim}{\rho} = \underset{\sim}{\rho}\underset{\sim}{A}^{-1}$, $\underset{\sim}{A}\epsilon G$, then the covariant exterior derivative

(2.3.29) $D\underset{\sim}{\rho} = d\underset{\sim}{\rho} - (-1)^k \underset{\sim}{\rho} \wedge \underset{\sim}{\Gamma}$

has the induced transformation law

(2.3.30) $'D('\underset{\sim}{\rho}) = (D\underset{\sim}{\rho}) \underset{\sim}{A}^{-1}$.

Finally if $\underset{\sim}{\Sigma}\epsilon\Lambda^k$ transforms according to the adjoint action of the group G : $'\underset{\sim}{\Sigma} = \underset{\sim}{A}\underset{\sim}{\Sigma}\underset{\sim}{A}^{-1}$, $\underset{\sim}{A}\epsilon G$, then

(2.3.31) $D\underset{\sim}{\Sigma} = d\underset{\sim}{\Sigma} + \underset{\sim}{\Gamma} \wedge \underset{\sim}{\Sigma} - (-1)^k \underset{\sim}{\Sigma} \wedge \underset{\sim}{\Gamma}$

has the transformation law

(2.3.32) $'D('\underset{\sim}{\Sigma}) = \underset{\sim}{A}(D\underset{\sim}{\Sigma})\underset{\sim}{A}^{-1}$.

From (2.3.27), (2.3.29), and (2.3.31), simple calculations show that

(2.3.33) $DD\underset{\sim}{\eta} = \underset{\sim}{\Theta} \wedge \underset{\sim}{\eta}$,

(2.3.34) $DD\underset{\sim}{\rho} = - \underset{\sim}{\rho} \wedge \underset{\sim}{\Theta}$

and

(2.3.35) $DD\underset{\sim}{\Sigma} = \underset{\sim}{\Theta} \wedge \underset{\sim}{\Sigma} - \underset{\sim}{\Sigma} \wedge \underset{\sim}{\Theta}$

where

$$\underset{\sim}{\Theta} = d\underset{\sim}{\Gamma} + \underset{\sim}{\Gamma} \wedge \underset{\sim}{\Gamma}$$

and $\underset{\sim}{\eta}$, $\underset{\sim}{\rho}$ and $\underset{\sim}{\Sigma}$ have the transformation laws

$'\eta = \underset{\sim}{A}\underset{\sim}{\eta}$, $'\underset{\sim}{\rho} = \underset{\sim}{\rho}\underset{\sim}{A}^{-1}$, $'\underset{\sim}{\Sigma} = \underset{\sim}{A}\underset{\sim}{\Sigma}\underset{\sim}{A}^{-1}$.

2.4 Antiexact Yang-Mills Fields

It is well known that the problem of solving Maxwell's equations is significantly simplified by an appropriate choice of gauge. Argument by analogy with electrodynamics leads directly to a similar expectation in the case of Yang-Mills fields. This expectation can be fully realized, and in a strictly algebraic way, as as we now proceed to show. Here, we follow the argument given in [13], Section 4.

Let G be an r-parameter matrix Lie group and, as before, let $\{\underset{\sim}{\gamma}_\alpha, \alpha = 1,\ldots,r\}$ be a basis for the Lie algebra of G. We saw in the last Section that any matrix $\underset{\sim}{\Gamma}$ of connection 1-forms of G takes values in the Lie algebra of G. Thus, any such $\underset{\sim}{\Gamma}$ belongs to the collection

$$(2.4.1) \qquad \mathcal{Y} = \{W^\alpha \underset{\sim}{\gamma}_\alpha \mid W^\alpha \in \Lambda^1(E_4) \ , \ \alpha = 1,\ldots,r\} \ .$$

In fact, at this point, any element of $\mathcal{Y}$ could act as a matrix of connection 1-forms of G. Further, it follows from (2.3.9) and (2.3.19) that any element $\underset{\sim}{A}$ of G generates the transformation

$$(2.4.2) \qquad {}'\underset{\sim}{\Gamma} = \underset{\sim}{A}\,\underset{\sim}{\Gamma}\,\underset{\sim}{A}^{-1} - d\underset{\sim}{A}\,\underset{\sim}{A}^{-1} \ .$$

and hence G may be considered to act on the collection $\mathcal{Y}$ via (2.4.2):

(2.4.3) $\quad G_Y\colon \mathcal{Y} \to \mathcal{Y} \mid {}'\underset{\sim}{\Gamma} = \underset{\sim}{A}\,\underset{\sim}{\Gamma}\,\underset{\sim}{A}^{-1} - d\underset{\sim}{A}\,\underset{\sim}{A}^{-1}\ , \quad \underset{\sim}{A} \in G\ .$

Since the Lagrangian function for the problem is invariant under the action of G (is gauge invariant), the Euler-Lagrange field equations for the problem are gauge covariant. Accordingly, the gauge group will map solutions of the field equations onto solutions, and hence the action of G_Y is an equivalence relation on $\mathcal{Y}$ (gauge equivalence). Under these circumstances, $\mathcal{Y}$ becomes a fiber space $\mathcal{Y}/G_Y$ under the identification of its equivalent elements. Thus, it is only necessary that we characterize a cross section K of this fiber space in order to know all of $\mathcal{Y}$, for $\mathcal{Y}$ is then the orbit of K under the action of G_Y .

Pick a point P of E_4 as center for the construction of the linear homotopy operator H , and let $\underset{\sim}{\Gamma}$ denote a generic element of $\mathcal{Y}$. The matrix $\underset{\sim}{A}_\Gamma$ that solves the Riemann-Graves integral equation

(2.4.5) $\quad \underset{\sim}{A}_\Gamma = \underset{\sim}{I} + H(\underset{\sim}{A}_\Gamma \underset{\sim}{\Gamma})$

belongs to G because $\underset{\sim}{\Gamma}$ is a matrix of infinitesimal generating 1-forms of G (see [3], p. 458ff). Exterior differentiation of (2.4.5) yields $d\underset{\sim}{A}_\Gamma = dH(\underset{\sim}{A}_\Gamma \underset{\sim}{\Gamma})$. However, H satisfies $dH + Hd$ = identity, and hence we have

$$(2.4.6) \qquad d\underset{\sim}{A}_\Gamma = \underset{\sim}{A}_\Gamma \underset{\sim}{\Gamma} - Hd(\underset{\sim}{A}_\Gamma \underset{\sim}{\Gamma}) .$$

When this is substituted into the right-hand side of (2.4.2), we obtain

$$'\underset{\sim}{\Gamma} = (Hd(\underset{\sim}{A}_\Gamma \underset{\sim}{\Gamma}))\underset{\sim}{A}_\Gamma^{-1} .$$

$'\underset{\sim}{\Gamma}$ thus belongs to the antiexact cross section

$$(2.4.7) \qquad A_Y = \{W^\alpha \gamma_\alpha \mid W^\alpha \in A^1(E_4) , \alpha = 1,\ldots,r\}$$

because $Hd(\underset{\sim}{A}_\Gamma \underset{\sim}{\Gamma})$ belongs to A_Y and the module property of antiexact forms. Thus, *any element of* Y *can be mapped onto an element of* A_Y *by an appropriate choice of an element of* G ; we have the map

$$(2.4.8) \qquad \pi: Y \to A_Y \mid {}'\underset{\sim}{\Gamma} = (\underset{\sim}{A}_\Gamma \underset{\sim}{\Gamma} - d\underset{\sim}{A}_\Gamma)\underset{\sim}{A}_\Gamma^{-1} ,$$

$$\underset{\sim}{A}_\Gamma = \underset{\sim}{I} + H(A_\Gamma \underset{\sim}{\Gamma}) .$$

Let $\underset{\sim}{A}_1$ and $\underset{\sim}{A}_2$ be the elements of G that map $\underset{\sim}{\Gamma}_1$ and $\underset{\sim}{\Gamma}_2$ onto the corresponding antiexact elements $'\underset{\sim}{\Gamma}_1$ and $'\underset{\sim}{\Gamma}_2$, respectively:

$$'\underset{\sim}{\Gamma}_1 = (\underset{\sim}{A}_1 \underset{\sim}{\Gamma}_1 - d\underset{\sim}{A}_1)\underset{\sim}{A}_1^{-1} , \quad '\underset{\sim}{\Gamma}_2 = (\underset{\sim}{A}_2 \underset{\sim}{\Gamma}_2 - d\underset{\sim}{A}_2)\underset{\sim}{A}_2^{-1} .$$

We thus have

(2.4.9) $\underset{\sim}{\Gamma}_1 = \underset{\sim}{A}_1^{-1}({}'\underset{\sim}{\Gamma}_1\underset{\sim}{A}_1 + d\underset{\sim}{A}_1)$, $\underset{\sim}{\Gamma}_2 = \underset{\sim}{A}_2^{-1}({}'\underset{\sim}{\Gamma}_2\underset{\sim}{A}_2 + d\underset{\sim}{A}_2)$.

Now, $\underset{\sim}{\Gamma}_1$ can be mapped onto $\underset{\sim}{\Gamma}_2$ by an element $\underset{\sim}{B}$ of G if and only if $\underset{\sim}{B}$ satisfies $\underset{\sim}{\Gamma}_2 = (\underset{\sim}{B}\underset{\sim}{\Gamma}_1 - d\underset{\sim}{B})\underset{\sim}{B}^{-1}$; that is, if and only if

(2.4.10) $d\underset{\sim}{B} = \underset{\sim}{B}\underset{\sim}{\Gamma}_1 - \underset{\sim}{\Gamma}_2\underset{\sim}{B}$.

If we make the substitution $\underset{\sim}{B} = \underset{\sim}{A}_2^{-1}\,\underset{\sim}{C}\,\underset{\sim}{A}_1$ and use (2.4.9) to eliminate $\underset{\sim}{\Gamma}_1$ and $\underset{\sim}{\Gamma}_2$ from (2.4.10), we obtain the requirement

(2.4.11) $d\underset{\sim}{C} = \underset{\sim}{C}{}'\underset{\sim}{\Gamma}_1 - {}'\underset{\sim}{\Gamma}_2\underset{\sim}{C}$.

Since ${}'\underset{\sim}{\Gamma}_1$ and ${}'\underset{\sim}{\Gamma}_2$ are antiexact, an integration of (2.4.11) by use of the homotopy operator H then gives

(2.4.12) $\underset{\sim}{C} = \underset{\sim}{K} + H(\underset{\sim}{C}{}'\underset{\sim}{\Gamma}_1 - {}'\underset{\sim}{\Gamma}_2\underset{\sim}{C}) = \underset{\sim}{K}$, $d\underset{\sim}{K} = \underset{\sim}{0}$,

where $\underset{\sim}{K}$ is a constant element of G (i.e., $\underset{\sim}{K}$ belongs to G_o) . When this result is put back into (2.4.11), we obtain the direct results

(2.4.13) $\underset{\sim}{K}{}'\underset{\sim}{\Gamma}_1 = {}'\underset{\sim}{\Gamma}_2\underset{\sim}{K}$, $\underset{\sim}{B} = \underset{\sim}{A}_2^{-1}\underset{\sim}{K}\underset{\sim}{A}_1$, $\underset{\sim}{K} \in \underset{\sim}{G}_o$.

Two elements of Y can be mapped onto each other by an

element of G *if and only if their images in* A_Y *are related through the adjoint action of an element of* G_o Thus, A_Y is unique to within the adjoint action of the original group G_o .

When an element $\underset{\sim}{\Gamma}$ of $\mathcal{Y}$ is mapped onto A_Y by an appropriate element of G , it becomes a Lie algebra-valued antiexact 1-form, $\underset{\sim}{\Gamma}_a$. Accordingly, $\underset{\sim}{\Gamma}_a$ satisfies the conditions

$$X \lrcorner \underset{\sim}{\Gamma}_a = \underset{\sim}{0} , \quad \underset{\sim}{\Gamma}_a(X_o^b) = \underset{\sim}{0} .$$

This process thus imposes a set of gauge conditions, *the antiexact gauge conditions*. The mapping π of $\mathcal{Y}$ onto A_Y , that is given by (2.4.8), may thus be viewed as restriction to the *antiexact gauge* cross section of $\mathcal{Y}$. The result of the last paragraph show that the antiexact gauge restriction is unique to within the adjoint action of the original group G_o . Accordingly, the antiexact gauge fixes $\underset{\sim}{\Gamma}$ to within the adjoint action of the original homogeneous group G_o . This result should not be a surprise, for without it an assignment of gauge would break the action of the original homogeneous group G_o of internal symmetries.

Up to this point, we have assumed that the center of E_4 has been fixed. Let H_1 and H_2 denote the linear homotopy operators that are constructed from two choices of the center of E_4, and let $\mathcal{A}_{Y1}$ and $\mathcal{A}_{Y2}$ be the corresponding antiexact cross sections of $\mathcal{Y}$ that are determined in the manner described above. If $\underset{\sim}{\Gamma}$ is a generic element of $\mathcal{Y}$, then

$$(2.4.14) \quad \underset{\sim}{A}_1 = \underset{\sim}{I} + H_1(\underset{\sim}{A}_1\underset{\sim}{\Gamma}) , \quad \underset{\sim}{A}_2 = \underset{\sim}{I} + H(\underset{\sim}{A}_2\underset{\sim}{\Gamma})$$

map $\underset{\sim}{\Gamma}$ onto the corresponding elements

$$(2.4.15) \quad \begin{aligned} {}'\underset{\sim}{\Gamma}_1 &= (\underset{\sim}{A}_1\underset{\sim}{\Gamma} - d\underset{\sim}{A}_1)\underset{\sim}{A}_1^{-1} \in \mathcal{A}_{Y1} , \\ {}'\underset{\sim}{\Gamma}_2 &= (\underset{\sim}{A}_2\underset{\sim}{\Gamma} - d\underset{\sim}{A}_2)\underset{\sim}{A}_2^{-1} \in \mathcal{A}_{Y2} , \end{aligned}$$

respectively. An elimination of the common term $\underset{\sim}{\Gamma}$ between these two equations then gives

$$(2.4.16) \quad {}'\underset{\sim}{\Gamma}_2 = (\underset{\sim}{B}\,{}'\underset{\sim}{\Gamma}_1 - d\underset{\sim}{B})\underset{\sim}{B}^{-1} , \quad \underset{\sim}{B} = \underset{\sim}{A}_2\underset{\sim}{A}_1^{-1} \in G .$$

Cross sections of $\mathcal{Y}$ *that are determined by different choices of the center of* E_4 *can be mapped one onto the other by the action of the gauge group* G . Thus, gauge covariance of the theory renders the theory insensative to the choice of center!

This latter result is of particular importance in practice, for it allows us to choose the center of E_4 in such a way as to give maximal simplification in actual calculations. An obvious choice is to take the origin of the coordinate cover as the center, in which case $X_o^a = 0$. We will see later that the choice of center actually corresponds to the choice of a reference point for the Yang-Mills potential functions $W_a^\alpha(X^b)$, while $W_a^\alpha(X^b)dX^a \in A^1(E_4)$ fixes the reference values at the center to be all zero (recall that $\omega \in A^1(E_4)$ implies $\omega(X_o^b) = 0$) .

2.5 Complete Exterior Systems of Equations

Complete exterior systems of equations play an important role in the development of the theory of defects. A one-to-one correspondence will be established between complete exterior systems and the kinematic field equations of defects of the solid state that are given in Section 2.7. Thus, the gauge transformation properties and representations of complete exterior systems lead to immediate similar information concerning the kinematics of defects of the solid state.

A collection of exterior forms $\{\underset{\sim}{\nu}, \underset{\sim}{\Gamma}, \underset{\sim}{\Sigma}, \underset{\sim}{\Theta}\}$ with $\underset{\sim}{\nu} \varepsilon \Lambda^{r}_{k,1}(E_n)$, $\underset{\sim}{\Gamma} \varepsilon \Lambda^{1}_{k,k}(E_n)$, $\underset{\sim}{\Sigma} \varepsilon \Lambda^{r+1}_{k,1}(E\ )$, $\underset{\sim}{\Theta} \varepsilon \Lambda^{2}_{k,k}(E_n)$, forms a complete differential system of degree r and class k on a starshaped region $S \subseteq E_n$ if and only if [3] (the $\underset{\sim}{\Gamma}$ and $\underset{\sim}{\Theta}$ used here are negatives of those appearing in the reference):

$$
\begin{aligned}
d\underset{\sim}{\nu} &= -\underset{\sim}{\Gamma} \wedge \underset{\sim}{\nu} + \underset{\sim}{\Sigma} \\
d\underset{\sim}{\Sigma} &= -\underset{\sim}{\Gamma} \wedge \underset{\sim}{\Sigma} + \underset{\sim}{\Theta} \wedge \underset{\sim}{\nu} \\
d\underset{\sim}{\Gamma} &= -\underset{\sim}{\Gamma} \wedge \underset{\sim}{\Gamma} + \underset{\sim}{\Theta} \\
d\underset{\sim}{\Theta} &= \underset{\sim}{\Theta} \wedge \underset{\sim}{\Gamma} - \underset{\sim}{\Gamma} \wedge \underset{\sim}{\Theta}
\end{aligned}
\tag{2.5.1}
$$

are satisfied throughout S . If $\underset{\sim}{\nu}$ and $\underset{\sim}{\Gamma}$ are given, the first of (2.5.1) serves to define $\underset{\sim}{\Sigma}$ and the third serves to define $\underset{\sim}{\Theta}$, in which case the second and fourth are identically satisfied. On the other hand, if $\underset{\sim}{\Sigma}$ and $\underset{\sim}{\Theta}$ are given, then the second and fourth of (2.5.1) constitute the integrability conditions for the first and third. It is in this sense that the system (2.5.1) is complete; all integrability conditions entailed by the system are contained in the system, the system is differentially closed. If $r = 1$, the equations (2.5.1) are the Cartan structure equations with $\underset{\sim}{\Gamma}$ being the connection 1-forms, $\underset{\sim}{\Theta}$ the curvature 2-forms, $\underset{\sim}{\Sigma}$ the torsion 2-forms, while $\underset{\sim}{\nu}$ are usually referred to as the soldering 1-forms. We adopt this same terminology here although it is usually used only in the context of a principal bundle that is a subbundle of the linear frame bundle.

It was noted in Section 2.2 that the linear homotopy operator H on a starshaped region S is the inverse of the exterior derivative on the module of antiexact differential forms on S . This fact has been used in [3] to effect an "integration" of complete exterior system that leads to the following representation:

$$(2.5.2) \qquad \underset{\sim}{\nu} = \underset{\sim}{A}\{d\underset{\sim}{\chi} + \underset{\sim}{\eta} - H(\underset{\sim}{\Gamma}_a \wedge d\underset{\approx}{\chi})\} ,$$

(2.5.3) $\underset{\sim}{\Sigma} = \underset{\sim}{A}\{d\underset{\sim}{\eta} + \underset{\sim}{\Gamma}_a \wedge \underset{\sim}{\eta} + H(d\underset{\sim}{\Gamma}_a \wedge d\underset{\sim}{\chi}) - \underset{\sim}{\Gamma}_a \wedge H(\underset{\sim}{\Gamma}_a \wedge d\underset{\sim}{\chi})\}$

(2.5.4) $\underset{\sim}{\Gamma} = \underset{\sim}{A}\underset{\sim}{\Gamma}_a\underset{\sim}{A}^{-1} - (d\underset{\sim}{A})\underset{\sim}{A}^{-1}$

(2.5.5) $\underset{\sim}{\Theta} = \underset{\sim}{A}(d\underset{\sim}{\Gamma}_a + \underset{\sim}{\Gamma}_a \wedge \underset{\sim}{\Gamma}_a)\underset{\sim}{A}^{-1}$

with $\underset{\sim}{A}\varepsilon\Lambda^{o}_{k,k}(E_n)$ the associated nonsingular *attitude* matrix; $\underset{\sim}{\chi}\varepsilon\Lambda^{r-1}_{k,1}(E_n)$ a column matrix of (r-1)-forms; $\underset{\sim}{\eta}\varepsilon A^{r}_{k,1}$ a column matrix of antiexact r-forms; and $\underset{\sim}{\Gamma}_a\varepsilon A^{1}_{k,k}$ a k-by-k square matrix of antiexact 1-forms. Furthermore, $\underset{\sim}{A}$ is the solution of the linear Riemann-Graves matrix integral equation

(2.5.6) $\underset{\sim}{A} = \underset{\sim}{I} - H(\underset{\sim}{\Gamma}\underset{\sim}{A})$,

and $\underset{\sim}{\chi}$, $\underset{\sim}{\eta}$ and $\underset{\sim}{\Gamma}_a$ are determined by

(2.5.7) $\underset{\sim}{\chi} = \underset{\sim}{\chi}_o + H(\underset{\sim}{A}^{-1}\underset{\sim}{\nu})$,

(2.5.8) $\underset{\sim}{\eta} = H(\underset{\sim}{A}^{-1}\underset{\sim}{\Sigma})$,

(2.5.9) $\underset{\sim}{\Gamma}_a = H(\underset{\sim}{A}^{-1}\underset{\sim}{\Theta}\underset{\sim}{A})$.

The attitude matrix of a complete exterior system is obtained by solving the Riemann-Graves integral equation (2.5.6), and is thus uniquely determined by $\underset{\sim}{\Gamma}$. On the other hand, there is an immediate resemblance between (2.5.4) and (2.4.2) with the exception that the $\underset{\sim}{\Gamma}$ in (2.4.2) takes its values in the Lie algebra of the gauge group G . Suppose, therefore, that the $\underset{\sim}{\Gamma}$ of a complete exterior system takes its values in the Lie algebra of a gauge group G . In this event, the solution of the Reimann-Graves integral equation (2.5.6) will yield an attitude matrix that belongs to G . Assume that G acts on $\underset{\sim}{\nu}$ and $\underset{\sim}{\Gamma}$ through the relations

$$(2.5.10) \quad {}'\underset{\sim}{\nu} = \underset{\sim}{A}\,\underset{\sim}{\nu}\ , \quad {}'\underset{\sim}{\Gamma} = (\underset{\sim}{A}\underset{\sim}{\Gamma} - d\underset{\sim}{A})\underset{\sim}{A}^{-1}\ , \quad \underset{\sim}{A} \in G$$

(compare (2.3.6) and (2.4.2). Then (2.5.1) implies

$$(2.5.11) \quad {}'\underset{\sim}{\Sigma} = \underset{\sim}{A}\underset{\sim}{\Sigma}\ , \quad {}'\underset{\sim}{\Theta} = \underset{\sim}{A}\ \underset{\sim}{\Theta}\ \underset{\sim}{A}^{-1}$$

(compare (2.4.20)). Under these circumstances, *the differential system* (2.5.1) *is covariant under the action of* G *(is gauge covariant) whenever* (2.5.10) *hold and* $\underset{\sim}{\Gamma}$ *takes its values in the Lie algebra of* G .

Complete exterior systems will be used in two ways in the theory to be constructed in the next Chapter. In the first instance, it will be used to obtain a general representation of the solutions of the kinematic field equations of defect dynamics. In this aspect, there is as yet no association of a gauge group and the quantities $\underset{\sim}{\Gamma}$ (and hence $\underset{\sim}{\Theta}$) will remain undetermined. The second instance obtains by showing that the minimal replacement construct associated with the underlying gauge group of elastically deformable solid bodies leads naturally to quantities $\underset{\sim}{\nu}$ and $\underset{\sim}{\Gamma}$ that form a complete exterior differential system for which (2.5.10) hold and for which $\underset{\sim}{\Gamma}$ takes its values in the Lie algebra of the gauge group. The gauge covariance of complete exterior systems then provides a mechanism for simplification and representation of the forms $\underset{\sim}{\nu}$, $\underset{\sim}{\Gamma}$, $\underset{\sim}{\Sigma}$, and $\underset{\sim}{\Theta}$ that is a generalization of the mapping π of Y onto A_Y considered in the previous Section. This result allows the investigator to impose the antiexact gauge. This significantly simplifies the considerations as well as conforming to the requirement that $\underset{\sim}{\Gamma}$ be antiexact that is dictated by the identification problem associated with the kinematics of defect dynamics. The reader should carefully note that (2.5.4) and (2.5.6) are the same as (2.4.2) and (2.4.5), for $'\underset{\sim}{\Gamma}$ in (2.4.2) corresponds to $\underset{\sim}{\Gamma}_a$ in (2.5.4) and hence $A_{\underset{\sim}{\Gamma}}$ in (2.4.5) coincides with the inverse of the matrix $\underset{\sim}{A}$ that appears in (2.5.7).

2.6 The Lagrangian of Elasticity Theory and Its Natural Gauge Group

The known system of fields considered in this study are those that characterize the solid state as a deformable continuum on the classic level. The starting point from which the rest will follow is therefore the Lagrangian function of the deformable elastic continuum.

Let E_3 denote 3-dimensional Euclidean space referred to a global Cartesian coordinate system (X^1, X^2, X^3). This space is taken to be the space of reference configurations of elastic bodies. An arc-wise connected, simply connected set B of nonzero Euclidean volume measure that is contained in a starshaped region S of E_3 is a reference configuration of an elastic body. Let $'E_3$ be a replica of E_3 that is referred to a global Cartesian coordinate system (x^1, x^2, x^3). A history of an elastic body is a diffeomorphism

$$\chi : B \times [T_o, T_1] \longrightarrow {}'E_3 \times [T_o, T_1] \mid$$

$$x^i = \chi^i(X^A, T),\ T = T$$

that renders the action functional

$$A[\chi] = \int_{T_o}^{T_1} \int_{\mathcal{B}} L_o(X^A, T, \chi^i, \partial_A \chi^i, \partial_4 \chi^i) dX^1 \wedge dX^2 \wedge dX^3 \wedge dT$$

$$= \int_{T_o}^{T_1} \int_{\mathcal{B}} L_o(X^A, T, \underset{\sim}{\chi}, \partial_A \underset{\sim}{\chi}, \partial_4 \underset{\sim}{\chi}) \ \mu \wedge dT$$

stationary in value relative to all competing diffeomorphisms that satisfy the same Dirichlet data, where the Lagrangian function L_o satisfies the invariance conditions

$$L_o(X^A, T+a, \underset{\sim}{A}\underset{\sim}{\chi}+\underset{\sim}{b}, \underset{\sim}{A}\partial_B \chi, \underset{\sim}{A}\partial_4 \underset{\sim}{\chi}) = L_o(X^A, T, \underset{\sim}{\chi}, \partial_B \underset{\sim}{\chi}, \partial_4 \underset{\sim}{\chi})$$

for all $a \in \mathbb{R}$, all $\underset{\sim}{b}$ such that $d\underset{\sim}{b} = \underset{\sim}{0}$, and all $\underset{\sim}{A}$ such that $\underset{\sim}{A}^T \underset{\sim}{A} = \underset{\sim}{I}$, $d\underset{\sim}{A} = \underset{\sim}{0}$. The kinetics of elastic bodies arises directly from the stated variational principle and invariance conditions via Noether's theorem [40, pp.176-179], and leads to the classic theory of nonlinear continua. On the other hand, the kinematics arises solely from the existence of the diffeomorphism χ and the implied continuity and differentiability properties. In fact, if we set $b^i = d\chi^i = \partial_a \chi^i \, dX^a$, then $db^i = 0$ and $b^1 \wedge b^2 \wedge b^3 \big|_{T=T^*} \neq 0$. Conversely,

$d\ell^i = 0$ implies $\ell^i = dH\ell^i$ (see §2.2) and hence $\ell^i = d\chi^i$ with $\chi^i = H\ell^i + k^i$, $\ell^1 \wedge \ell^2 \wedge \ell^3 \big|_{T=T^*} = \left.\dfrac{\partial(\chi^1,\chi^2,\chi^3)}{\partial(X^1,X^2,X^3)}\right|_{T=T^*} \neq 0$. Thus, the existence of the diffeomorphism χ implies and is implied by the existence of 3 1-forms $\{\ell^1,\ell^2,\ell^3\}$ such that $d\ell^i = 0$, $\ell^1 \wedge \ell^2 \wedge \ell^3 \big|_{T=T^*} \neq 0$.

The Lagrangian that describes elasticity theory is of the form

$$(2.6.1) \qquad L_o = T - \Psi(C_{AB}) \ .$$

where T denotes the kinetic energy

$$(2.6.2) \qquad T = \frac{1}{2}\,\rho\,\partial_4\chi^i\,\delta_{ij}\,\partial_4\chi^j \ ,$$

and $\Psi(C_{AB})$ is the potential energy that is a function of the Cauchy strain tensor whose components are given by

$$(2.6.3) \qquad C_{AB} = \partial_A\chi^i\,\delta_{ij}\,\partial_B\chi^j \ .$$

If we define components of a 4-dimensional tensor $\underset{\sim}{C}$ by

$$(2.6.4) \qquad C_{ab} = \partial_a\chi^i\,\delta_{ij}\,\partial_b\chi^j \ , \quad C_{ab} = C_{ba} \ ,$$

then the Lagrangian L_o is a function of C_{ab}, i.e.

(2.6.5) $$L_o = L_o(C_{ab}) = T(C_{44}) - \Psi(C_{AB}) .$$

The underlying group G_o for elasticity theory is a semi-direct product, $G_o = SO(3)_o \triangleright T(3)_o$, of the proper real orthogonal group $SO(3)_o$ and the translation group $T(3)_o$. We proceed to show that it is an invariance group for the Lagrangian (2.6.5).

The action of the group G_o on a state vector $\underset{\sim}{\chi}$ is realized by

(2.6.6) $$'\underset{\sim}{\chi} = \underset{\sim}{A}\underset{\sim}{\chi} + \underset{\sim}{b} \qquad \underset{\sim}{A} \varepsilon SO(3)_o , \quad \underset{\sim}{b} \varepsilon T(3)_o$$

where $\underset{\sim}{A}$ is an orthogonal constant matrix and $\underset{\sim}{b}$ is a constant column vector. The expression (2.6.4) can be written in an equivalent way

(2.6.7) $$C_{ab} = \partial_a \underset{\sim}{\chi}^T \partial_b \underset{\sim}{\chi} .$$

Thus, under the action of the homogeneous group $G_o = SO(3)_o \triangleright T(3)_o$, namely under the transformation (2.6.6) with $\underset{\sim}{A}\underset{\sim}{A}^T = \underset{\sim}{I}$, $d\underset{\sim}{A} = \underset{\sim}{0}$ and $d\underset{\sim}{b} = \underset{\sim}{0}$, the components of the tensor $\underset{\sim}{C}$ transform according to

(2.6.8) $$'C_{ab} = \partial_a {'\underset{\sim}{\chi}}^T \partial_b {'\underset{\sim}{\chi}} = \partial_a \underset{\sim}{\chi}^T \underset{\sim}{A}^T \underset{\sim}{A} \partial_b \underset{\sim}{\chi} = C_{ab} .$$

Hence, the Lagrangian (2.6.5) is invariant under the homogeneous action of the group G_o

(2.6.9) $$L_o('C_{ab}) = L_o(C_{ab}) \; .$$

The group G_o is thus an invariance group for the Lagrangian L_o . Each of the homogeneous subgroups, $SO(3)_o$ and $T(3)_o$, of the group G_o is also an invariance group of L_o .

2.7 The Kinematics of Defects

The physics of elastic bodies with defects (dislocations, disclinations, voids, inclusions, etc.) is obtained through consideration of current configurations that can not be characterized solely in terms of a diffeomorphism χ from a reference configuration. The simplest way of stating this is to replace the requirements $db^i = 0$, $b^1 \wedge b^2 \wedge b^3 \big|_{T=T^*} \neq 0$ by

$$dB^i \neq 0 \ , \quad B^1 \wedge B^2 \wedge B^3 \big|_{T=T^*} \neq 0 \ .$$

If we write $B^i = B^i_a \, dX^a$, we have $B^i = dHB^i + HdB^i$ and hence

$$B^i = d\chi^i + HdB^i \ , \quad \chi^i = HB^i + k^i \ .$$

Thus, the independent 1-forms B^i give rise to a completely integrable part, $d\chi^i$, and to an antiexact (non-integrable) part HdB^i that is determined by the 2-forms dB^i . Clearly, the part HdB^i characterizes the internal degrees of freedom of the material configuration; that is, defects are obstructions to the existence of diffeomorphisms χ that uniquely characterize current configurations of a body.

There are two fundamental questions involved here. The first is the transition from the exact independent 1-forms b^i to the nonexact but independent 1-forms B^i. This is not a question of taste, rather it is an inescapable necessity when the action of the group $SO(3)_o \triangleright T(3)_o$ is allowed to become inhomogeneous while preserving the invariance of the Lagrangian function. The substance of the argument involved here is the Yang-Mills construct that was summarized in Section 2.3. The second is the meaning and interpretation of a current state of a body that does not arise solely from a smooth mapping from a reference configuration. This latter question will be discussed in Section 3.7 in terms of the observables associated with material bodies with defects.

It is clear from the outset that a replacement of the classical kinematics of elastic continua , $db^i = 0$, $b^1 \wedge b^2 \wedge b^3\big|_{T=T^*} \neq 0$, by a kinematics with defects, $dB^i \neq 0$, $B^1 \wedge B^2 \wedge B^3\big|_{T=T^*} \neq 0$, entails a radical shift in the customary physical notions. There is no fundamental difficulty here, for one simply constructs a complete exterior system starting with the 1-forms B^i

as described in Section 2.5 (i.e., $B^i = \nu^i$ in (2.5.1)). This gives the Cartan equations of structure that are naturally associated with the 1-forms B^i. Hence connection 1-forms, torsion 2-forms and curvature 2-forms are naturally associated with states of bodies characterized by the 1-forms B^i. However, the mechanics of materials with defects has grown up in a context substantially different from that envisioned by E. Cartan; the kinematics was developed through analogies with elasticity theory and the theory of the plastic state. This has given rise to physically natural definitions of 1 and 2 point tensor fields such as dislocation and disclination densities and currents, spin, bend-twist, distortion and distortion velocity [10,17,18]. A full listing of these equations is given in this Section together with the currently accepted forms of the equations of balance of linear momentum for materials with defects. Consistency between the two approaches will be established in Section 3.1 where we show that the kinematic equations of defect dynamics can be placed in a 1-to-1 correspondence with the Cartan equations of structure.

The continuity equations of defect dynamics are given in terms of the 3-dimensional exterior calculus by [5]:

$$\partial_4\alpha^i = -\bar{d}J^i - S^i , \qquad \partial_4 Q^i = -\bar{d}S^i ,$$

(2.7.1)

$$\bar{d}\alpha^i = Q^i , \qquad \bar{d}Q^i = 0 ,$$

where $\left(\bar{d} = d\Big|_{T=\text{constant}}\right)$

(2.7.2)

$\alpha^i = \alpha^{Ai}\ \mu_A$ = 2-forms of dislocation density,

$J^i = J^i_A\ dX^A$ = 1-forms of dislocation current,

$S^i = S^{Ai}\ \mu_A$ = 2-forms of disclination current,

$Q^i = q^i\ \mu$ = 3-forms of disclination density,

and ∂_4 stands for the partial derivative with respect to time. Equations (2.7.1) can be written in terms of the components indicated in the equations (2.7.2). When this is done, we obtain the explicit set of equations:

(2.7.3)

$$\partial_4\alpha^{Ai} = -\ e^{ABC}\ \partial_B J^i_C - S^{Ai} ,$$

$$\partial_A\alpha^{Ai} = q^i ,$$

$$\partial_4 q^i = -\ \partial_A S^{Ai} ,$$

the equations $\bar{d}Q^i = 0$ being identically satisfied since $\bar{d}Q^i \epsilon \Lambda^4(E_3)$ and any 4-form on a 3-space vanishes identically.

The kinematic field equations of defects are the first integrals of the system (2.7.1) [5]. They are given by

$$\partial_4 k^i = \bar{d}\omega^i - S^i , \qquad \bar{d}k^i = Q^i ,$$
$$\partial_4 \beta^i = \bar{d}V^i - J^i - \omega^i , \qquad \bar{d}\beta^i = \alpha^i - k^i , \tag{2.7.4}$$

with

$$k^i = k^{Ai}\,\mu_A = \text{bend-twist 2-forms},$$
$$\omega^i = \omega^i_A\, dX^A = \text{spin 1-forms},$$
$$\beta^i = \beta^i_A\, dX^A = \text{distortion 1-forms}, \tag{2.7.5}$$
$$V^i = \text{velocity 0-forms} .$$

Equations (2.7.4) describe the internal state of a body with defects. When they are resolved on the basis elements $\{\mu_A, dX^A\}$, we obtain the possibly more familiar component representations [17,18]

$$\partial_4 k^{Ai} = - S^{Ai} + e^{ABC}\,\partial_B\omega^i_C , \quad \partial_A k^{Ai} = q^i ,$$
$$\partial_4 \beta^i_A = \partial_A V^i - J^i_A - \omega^i_A , \quad e^{ABC}\,\partial_B\beta^i_C = \alpha^{Ai} - k^{Ai} . \tag{2.7.6}$$

The formulation of the kinematic equations of defects given above is based on the assumption of an underlying space as the Cartesian product, $E_3 \times \mathbb{R}$, of a 3-dimensional Euclidean reference space, E_3 , and an independent time scale with $T \in \mathbb{R}$. Although this formulation is adequate, certain intrinsic structural relations and analogies are directly revealed by a true 4-dimensional space-time formulation of the basic equations (see Appendix 4).

Keeping in mind the results from section 2.1 on the exterior calculus, we can rewrite equations (2.7.1) and (2.7.4) in terms of quantities defined on a 4-dimensional space E_4 with coordinate cover $\{X^a\} = \{X^A, T\}$. The system of continuity equations (2.7.1) of defect dynamics are satisfied [5] if and only if the disclination 3-forms

(2.7.7) $$\Omega^i = - S^i \wedge dT + Q^i = - S^{Ai} \mu_A \wedge dT + q^i \mu$$

and the dislocation 2-forms

(2.7.8) $$\mathcal{D}^i = J^i \wedge dT + \alpha^i = J^i_A \, dX^A \wedge dT + \alpha^{Ai} \mu_A$$

satisfy the exterior differential equations

(2.7.9) $$d\Omega^i = 0 \ , \quad d\mathcal{D}^i = \Omega^i$$

throughout the region of E_4 that is occupied by the space-time history of the body. Further, the system of continuity equations (2.7.9) admits the system of first integrals

(2.7.10) $\mathcal{D}^i = dB^i + \mathcal{K}^i$

in terms of the velocity-distortion 1-forms

(2.7.11) $B^i = V^i \, dT + \beta^i = V^i \, dT + \beta^i_A \, dX^A = B^i_a \, dX^a$

and the spin-twist 2-forms

(2.7.12) $\mathcal{K}^i = - \omega^i \wedge dT + k^i = - \omega^i_A \, dX^A \wedge dT + k^{Ai} \, \mu_A$.

The theory is not complete unless the field equations listed thus far are supplemented by appropriate statements of the law of balance of the linear momentum. Let $\mathcal{T} = \mathcal{T}(B^i_4)$ be a kinetic energy and $\Psi = \Psi(B^i_A)$ a potential energy of the system which depends on the distortions. This is what has been done historically [10]: the dependence of the classic potential energy on the deformation gradients is replaced by an identical dependence on the distortions. The replacement of the deformation gradients by the distortion field at this point is arbitrary. There is nothing so far that would justify this move except the hand-waving argument that the integrable displacement gradients, $d\chi^i = \partial_A \chi^i \, dX^A$ should be replaced by the new quantities, distortions, that are nonintegrable, when defects are present. This assumption has its roots in the idea that a "plastic" distortion, $\overset{P}{B}{}^i_A$, should not give rise to a stress response and that $\partial_A \chi^i = B^i_A + \overset{P}{B}{}^i_A$. Clearly, these assumptions involve implied relations between defects and plasticity theory that have, as yet, not been clearly stated. The theory developed in the next chapter will show that $T = T(B^i_4)$ and $\Psi = \Psi(B^i_A)$ are correct from fundamental principles rather than an inexplicit appeal to plasticity theory.

We now proceed by defining the 3-forms Z_i by

(2.7.13) $$Z_i = \frac{\partial(T-\Psi)}{\partial B^i_a} \pi_a = -\frac{\partial \Psi}{\partial B^i_A} \pi_A + \frac{\partial T}{\partial B^i_4} \pi_4 .$$

The standard variational definitions identify

(2.7.14) $$p_i = \frac{\partial T}{\partial B^i_4}$$

as the components of linear momentum and

(2.7.15) $$\sigma^A_i = \frac{\partial \Psi}{\partial B^i_A}$$

as the components of the Piola-Kirchhoff stress tensor. The equations of balance of linear momentum are then given by

(2.7.16) $$dZ_i = (\partial_4 p_i - \partial_A \sigma^A_i)\pi = 0 .$$

The 3-forms Z_i can be written in the following way

(2.7.17) $Z_i = -\sigma_i^A \pi_A + p_i \pi_4 = -\sigma_i^A \mu_A \wedge dT - p_i \mu$

so that (2.7.16) follows from (2.1.9) and (2.1.10).

For later convenience we rewrite the equations (2.7.9), (2.7.10) and (2.7.16) in matrix notation.

(2.7.18) $d\underset{\sim}{\Omega} = \underset{\sim}{0}$, $\underset{\sim}{\Omega} = d\underset{\sim}{\mathcal{D}} = d\underset{\sim}{K}$

(2.7.19) $\underset{\sim}{\mathcal{D}} = d\underset{\sim}{B} + \underset{\sim}{K}$,

(2.7.20) $d\underset{\sim}{Z} = \underset{\sim}{0}$,

where $\underset{\sim}{B} \varepsilon \Lambda^1_{3,1}(E_4)$ is a column matrix whose entries are the 1-forms B^1 , B^2 , B^3 , $\underset{\sim}{K}, \underset{\sim}{\mathcal{D}} \varepsilon \Lambda^2_{3,1}(E_4)$ are column matrices whose entries $\{K^i\}$, $\{\mathcal{D}^i\}$ are 2-forms defined on E_4 , $\underset{\sim}{\Omega} \varepsilon \Lambda^3_{3,1}(E_4)$ is a column matrix of 3-forms on E_4 with entries $\{\Omega^i\}$, and $\underset{\sim}{Z} \varepsilon \Lambda^3_{1,3}(E_4)$ is a row matrix of 3-forms with entries $\{Z_i\}$.

Chapter 3

THE GAUGE THEORY OF DEFECTS

3.1 The Inverse and Cauchy Problems in Defect Dynamics

The continuum theory of defects summarized in Section 2.7 departs drastically from the customary formulations of continuum theories. Previous theories of the solid continuum assume continuity and differentiability of mappings from a reference state into a current state; that is $d\underset{\sim}{B} = \underset{\sim}{0}$, $\underset{\sim}{K} = \underset{\sim}{0}$, $\underset{\sim}{D} = \underset{\sim}{0}$, $\underset{\sim}{\Omega} = \underset{\sim}{0}$ in which case we have $\underset{\sim}{B} = d\underset{\sim}{\chi}$ where $\underset{\sim}{x} = \underset{\sim}{\chi}$ defines the current state in terms of the mapping functions $\underset{\sim}{\chi}$. With defects present, these familiar equations are replaced by (2.7.18), (2.7.19),

$$d\underset{\sim}{\Omega} = \underset{\sim}{0} \ , \quad \underset{\sim}{\Omega} = d\underset{\sim}{D} = d\underset{\sim}{K} \ , \quad \underset{\sim}{D} = d\underset{\sim}{B} + \underset{\sim}{K} \ .$$

It thus follows that there must be other quantities present in addition to $\underset{\sim}{\chi}$ in order to characterize the kinematics of states with defects. The first problem that must be solved is therefore that of obtaining a system of state variables that provides a complete discription of states with defects. Once such a system is known, the required constitutive theory and the resulting kinetics become rational and well posed problems.

Consider a complete system of exterior equations of the form given by (2.5.1) and set $r = 1$, $k = 3$, and $n = 4$. With these as the starting point, we make the identifications:

$$(3.1.1) \quad \underset{\sim}{B} = \underset{\sim}{\nu} \ , \quad \underset{\sim}{\mathcal{D}} = \underset{\sim}{\Sigma} \ ,$$

where $\underset{\sim}{B}$ and $\underset{\sim}{\mathcal{D}}$ are the matrices of distortion 1-forms and the dislocation 2-forms defined in Section 2.7. In view of equations (2.5.1), the Cartan structure equations, and (2.7.18-19), the choice

$$(3.1.2) \quad \underset{\sim}{\Omega} = \underset{\sim}{\Theta} \wedge \underset{\sim}{\nu} - \underset{\sim}{\Gamma} \wedge \underset{\sim}{\Sigma} = \underset{\sim}{\Theta} \wedge \underset{\sim}{B} - \underset{\sim}{\Gamma} \wedge \underset{\sim}{\mathcal{D}}$$

for the matrix of disclination 3-forms and

$$(3.1.3) \quad \underset{\sim}{K} = \underset{\sim}{\Gamma} \wedge \underset{\sim}{\nu} = \underset{\sim}{\Gamma} \wedge \underset{\sim}{B}$$

for the matrix of bend-twist 2-forms is such that equations (2.7.18-19) of defect dynamics are identically satisfied. Further, this identical satisfaction of the equations of defect dynamics obtains for any and every choice of the quantities $\underset{\sim}{\Gamma}$ and $\underset{\sim}{\Theta}$ since these quantities are, as yet, unidentified. Therefore, if we can find a general representation of the system (2.5.1), the kinematic equations of defect dynamics will be solved.

It follows directly from (2.5.2-3) and (3.1.1) that the identifications (3.1.1) are equivalent to

(3.1.4) $\quad \underset{\sim}{B} = \underset{\sim}{A}[d\underset{\sim}{\chi} + \underset{\sim}{\eta} - H(\underset{\sim}{\Gamma}_a \wedge d\underset{\sim}{\chi})]$

(3.1.5) $\quad \underset{\sim}{D} = \underset{\sim}{A}[d\underset{\sim}{\eta} + \underset{\sim}{\Gamma}_a \wedge \underset{\sim}{\eta} + H(d\underset{\sim}{\Gamma}_a \wedge d\underset{\sim}{\chi}) - \underset{\sim}{\Gamma}_a \wedge H(\underset{\sim}{\Gamma}_a \wedge d\underset{\sim}{\chi})]$.

A simple calculation and (3.1.2-3) then give

(3.1.6) $\quad \underset{\sim}{K} = \underset{\sim}{A}(\underset{\sim}{\Gamma}_a - \underset{\sim}{A}^{-1} d\underset{\sim}{A}) \wedge [d\underset{\sim}{\chi} + \underset{\sim}{\eta} - H(\underset{\sim}{\Gamma}_a \wedge d\underset{\sim}{\chi})]$

and

(3.1.7) $\quad \underset{\sim}{\Omega} = \underset{\sim}{A}\{(d\underset{\sim}{\Gamma}_a + \underset{\sim}{\Gamma}_a \wedge \underset{\sim}{\Gamma}_a) \wedge [d\underset{\sim}{\chi} + \underset{\sim}{\eta} - H(\underset{\sim}{\Gamma}_a \wedge d\underset{\sim}{\chi})] - (\underset{\sim}{\Gamma}_a$

$$- \underset{\sim}{A}^{-1} d\underset{\sim}{A}) \wedge [d\underset{\sim}{\eta} + \underset{\sim}{\Gamma}_a \wedge \underset{\sim}{\eta} + H(d\underset{\sim}{\Gamma}_a \wedge d\underset{\sim}{\chi}) - \underset{\sim}{\Gamma}_a \wedge H(\underset{\sim}{\Gamma}_a \wedge d\underset{\sim}{\chi})]\} .$$

The entries $\{\eta^i\}$ and $\{\Gamma^i_j\}$ of the matrices $\underset{\sim}{\eta}$ and $\underset{\sim}{\Gamma}_a$, respectively, are 1-forms and can be written in terms of a basis for the 4-dimensional vector space $\Lambda^1(E_4)$ as follows:

(3.1.8) $\quad \eta^i = \eta^i_b \, dX^b \quad \text{and} \quad \Gamma^i_j = \Gamma^i_{bj} \, dX^b$.

From (2.5.8-9) it follows that η^i and Γ^i_j are antiexact 1-forms and hence [see(2.2.5e)] the 12+36 functions η^i_b and Γ^i_{bj} must satisfy 3+9 algebraic conditions

(3.1.9) $\quad X^b \eta^i_b = 0$ and $X^b \Gamma^i_{bj} = 0 \quad (X^b_o = 0)$.

Equations (3.1.4-7) present explicit evaluation of all of the quantities of defect dynamics in terms of the $3 + (12-3) + (36-9) + 9$ functions χ^i , η^i_b , Γ^i_{bj} and A^i_j . The problem of solving the equations of defect dynamics is thus equivalent to the problem of determining these 48 functions.

Suppose, for the moment, that we could justify choosing $\underset{\sim}{A}(\underset{\sim}{X})$ to be the identity matrix, $\underset{\sim}{I}$. This would reduce the number of unknown functions to 39 . However, since the equations of defect dynamics (2.7.18-19) are now identically satisfied, we have left only the three equations (2.7.20) of the balance of linear momentum and these are clearly insufficient for the determination of the 39 unknowns $(\chi^i, \eta^i_b, \Gamma^i_{bj})$.

The underlying problem can be simplified further if we restrict our attention to disclination free materials. In this case we have $\underset{\sim}{\Omega} = \underset{\sim}{0}$, $\underset{\sim}{K} = \underset{\sim}{0}$, and, since we have taken $\underset{\sim}{A} = \underset{\sim}{I}$, (3.1.4-7) show that we must likewise have $\underset{\sim}{\Gamma}_a = \underset{\sim}{0}$ if $\underset{\sim}{B}$ is not to be exact (i.e. completely integrable). Thus, (3.1.4-5) give

(3.1.10) $\quad \underset{\sim}{B} = d\underset{\sim}{\chi} + \underset{\sim}{\eta}$ and $\underset{\sim}{D} = d\underset{\sim}{\eta}$

for disclination free materials. It is now easy to show

that the so called inverse problem is well posed:

> determine the distortion and velocity matrix $\underset{\sim}{B}$ of 1-forms for a given dislocation density and current matrix $\underset{\sim}{\mathcal{D}}$ of 2-forms that satisfies the continuity equation $d\underset{\sim}{\mathcal{D}} = \underset{\sim}{0}$.

The matrix $\underset{\sim}{\eta}$ is a matrix whose entries are antiexact 1-forms, and the homotopy operator H inverts the operator d of exterior differentiation on the module of antiexact forms [see(2.2.5)]. Thus, an application of the homotopy operator H to both sides of $(3.1.10)_2$ serves to determine $\underset{\sim}{\eta}$ and (3.1.10) becomes

(3.1.11) $\underset{\sim}{\eta} = H(\underset{\sim}{\mathcal{D}})$, $\underset{\sim}{B} = d\underset{\sim}{\chi} + H(\underset{\sim}{\mathcal{D}})$.

The three remaining unknowns, χ^i , can now be determined by the three equations of balance of linear momemtum, $dZ_i = 0$. Here, of course one uses the standard constitutive relations of dislocation dynamics that relate the momentum, p_i , and the stress, σ_i^A , to the entries of $\underset{\sim}{B}$, distortion β_A^i , and velocity, V^i . (Recall that by (2.7.16-17) $Z_i = -\sigma_i^A \pi_A + p_i \pi_4$ and that $dZ_i = 0$ are the balance of linear momentum equations $\partial_4 p_i - \partial_A \sigma_i^A = 0$.) Explicit solutions of a number of static inverse problems are reported in the literature [10,20,18] . For static problems, the method of Green's functions for the equilibrium equations $\partial_A \sigma_i^A = 0$ provides the vehicle most often used to obtain the explicit representation of the solution.

It should now be evident that, although the inverse problem is well posed, the Cauchy problem in defect dynamics is not; there are simply not enough equations present in order to determine all of the unknown functions that enter into the expressions given by (3.1.4-7). One might think that adding an equation of balance of energy and the customary practices of non-equilibrium thermodynamics would lead to a well posed Cauchy problem (Appendix 4). Although such an approach meets with partial success (see [5]) one also has to introduce either a temperature or an entropy function and a dissipation function. The energy equation will serve for the determination of the entropy or temperature variable, but this is no real help since the determination of the dissipation function as a function of the thermodynamic forces is equivalent to the problem of rendering the Cauchy problem well posed. Unfortunately, there is no direct guide to the selection of the dissipation function for materials with defects.

Our aim in this work is to provide a system of equations that is sufficient in number to determine all of the relevant variables from given Cauchy data.

3.2 The Reference Configuration, a Reexamination

Let us go back to equations (3.1.4-7). They will be our starting point since we know that the equations of defect dynamics are identically satisfied once we express $\underset{\sim}{B}$, $\underset{\sim}{\mathcal{D}}$, $\underset{\sim}{K}$ and $\underset{\sim}{\Omega}$ in terms of $\underset{\sim}{\chi}$, $\underset{\sim}{\eta}$, $\Gamma_{\sim a}$ and $\underset{\sim}{A}$. As mentioned in the previous section there are 48 functions that are required in order to determine $\underset{\sim}{\chi}$, $\Gamma_{\sim a}$, $\underset{\sim}{\eta}$ and $\underset{\sim}{A}$. For purpose of argument let us suppose that the 39 needed to determine, $\underset{\sim}{\chi}$, $\underset{\sim}{\eta}$, $\Gamma_{\sim a}$, have been given. In this event it remains to determine the nine entries of the matrix $\underset{\sim}{A}$.

The concept of a reference configuration is a fundamental aspect of classical continuum mechanics, although its essential properties are often glossed over. The very concepts of strain and deformation demand a reference configuration for their quantification, while the Piola-Kirchhoff stress is defined relative to the reference configuration for the measure of associated surface area. The reference configuration in classical elasticity is understood as a configuration that is stress free and strain free. However, when there are internal degrees of freedom present, such as those encountered in defect dynamics, it is necessary that a careful reexamination of the underlying concept of a reference configuration be made.

In order to maintain consistency with classical continuum mechanics, we require the reference configuration of a material body with internal degrees of freedom to be one that is

(a) strain free,

(b) stress free.

To these, we add the further requirement that it be

(c) defect free.

This latter requirement is necessary in order that the concept of deformation may be maintained.

A body without defects is characterized by an integrable response, $x^i = \chi^i(X^a)$, where $\{x^i\}$ are the coordinates of a material point in the current configuration that occupied a point with coordinates $\{X^A\}$ in the reference configuration. This situation is characterized in the theory of materials with defects by the requirements

$$\underset{\sim}{B} = d\underset{\sim}{\chi} \ , \quad \underset{\sim}{\mathcal{D}} = \underset{\sim}{K} = \underset{\sim}{0} \ , \quad \underset{\sim}{\Omega} = \underset{\sim}{0} \ .$$

Comparing these last expressions with (3.1.4), we conclude that the defect free state is chacterized by

$$\underset{\sim}{\Gamma}_a = \underset{\sim}{0} \ , \quad \underset{\sim}{\eta} = \underset{\sim}{0} \ , \quad \underset{\sim}{A} = \underset{\sim}{I} \ .$$

Indeed, unless $\underset{\sim}{A} = \underset{\sim}{I}$, equations (3.1.4) give

$$B^i(X^a) = A^i_j(X^a)\ d\chi^j(X^a) \tag{3.2.1}$$

in the absence of defects and the 1-forms $\{B^i\}$ cease to be characterized by the diffeomorphism

$$\chi : B\times[T_o,T_1] \to {}'E_3\times[T_o,T_1] \,|\, x^i = \chi^i(X^A,T),\ T=T \tag{3.2.2}$$

from the reference configuration of the body into the history of the current configuration. Although equations (3.2.1) imply the complete integrability of the 1-forms B^i by the Frobenius theorem [3,4], this is not enough if we demand a unique correlation between the kinematic conditions $dB^i = 0$ of a defect free body and the associated diffeomorphism (3.2.2). Reflection on these facts show that the choice of the matrix $\underset{\sim}{A}$ is equivalent to the choice of a representation for the field quantities B^i , D^i , K^i , Ω^i since the quantities $\underset{\sim}{\Gamma}_a$ and $\underset{\sim}{\eta}$ are as yet undetermined.

The representation that will be used from now on is that given by (3.1.4-7) *with* $\underset{\sim}{A} = \underset{\sim}{I}$.

The representation given by $\underset{\sim}{A} = \underset{\sim}{I}$ leads to the representation

$$B^i = d\chi^i = dx^i \tag{3.2.3}$$

in the absence of defects. Accordingly, the distortion-velocity 1-forms of a defect free body from a globally parallel basis for 1-forms in the space $'E_3$ of current configuration. This allows us to associate a diffeomorphism χ with any state of a body with defects by the prescription $\chi^i(X^a) = HB^i + k^i$ (i.e., by the exact part of B^i): *we identify the functions* $\chi^i(X^a)$ *with the deformation functions of the completely integrable part of the response.*

It will be shown in the next Section that the choice $\underset{\sim}{A} = \underset{\sim}{I}$ is equivalent to the demand that the connection matrix $\underset{\sim}{\Gamma}$ that appears in the Cartan equations of structure has antiexact 1-forms as entries. This is reminiscent of the considerations layed out in Section 2.4 whereby any $\underset{\sim}{\Gamma}$ in $\mathcal{Y}$ could be mapped onto an element of $\mathcal{A}_Y$ by the action of the gauge group. At this point, we anticipate later arguments (Section 3.6) that will show that $\underset{\sim}{\Gamma}$ takes its values in the Lie algebra of the gauge group $G = SO(3)>T(3)$ and that $\underset{\sim}{B}$ $(=\underset{\sim}{\nu})$ and $\underset{\sim}{\Gamma}$ transform under the action of G by (2.5.10). Accordingly, the Cartan structure equations (2.5.1) are gauge covariant and we may replace $\underset{\sim}{\Gamma}$ by $\underset{\sim}{\Gamma}_a$ with no loss of generality by choosing the antiexact gauge. Once we have found solutions with $\underset{\sim}{A} = \underset{\sim}{I}$, we can always perform a gauge transformation by allowing $'\underset{\sim}{A} \varepsilon G$ to act and thereby obtain solutions in different gauges. It must be noted, however, that the matrices $\underset{\sim}{\Gamma}_a$, $\underset{\sim}{\eta}$ and $\underset{\sim}{\chi}$ will not have the same physical meanings in these different gauges. In point of fact, it is only in the antiexact gauge that $\underset{\sim}{\chi}$ may be identified with the deformation functions associated with the current state; i.e. $\underset{\sim}{B} = \underset{\sim}{A}\{d\underset{\sim}{\chi} + \underset{\sim}{\eta} - H(\underset{\sim}{\Gamma}_a \wedge d\underset{\sim}{\chi})\}$ with $d\underset{\sim}{A} \neq \underset{\sim}{0}$ does not have $\underset{\sim}{A}\, d\underset{\sim}{\chi}$ as exact part.

Finally, we note that any attitude matrix $\underset{\sim}{A}$ satisfies the Riemann-Graves integral equation $\underset{\sim}{A} = \underset{\sim}{I} - H(\underset{\sim}{\Gamma}\underset{\sim}{A})$, and hence

$$\underset{\sim}{A}(X_0^a) = \underset{\sim}{I}$$

because H applied to anything followed by an evaluation at the center gives zero. Accordingly, the matrix $A^i_j(X^a)$ that appears in (3.2.1) is constant if and only if $\underset{\sim}{A} = \underset{\sim}{I}$; the B^i's are globally parallel in the current configuration in the absence of defects if and only if $\underset{\sim}{A} = \underset{\sim}{I}$.

3.3 The Connection Matrix

By the definition of exterior differentiation of the product of differential forms we have

(3.3.1) $d(\underset{\sim}{\Gamma}_a \underset{\sim}{\chi}) = d\underset{\sim}{\Gamma}_a \underset{\sim}{\chi} - \underset{\sim}{\Gamma}_a \wedge d\underset{\sim}{\chi}$.

The entries of the matrix $\underset{\sim}{\Gamma}_a$ are antiexact 1-forms [see(2.5.9)]. And, since the set $A(E_4)$ of all antiexact forms is a module, the product $\underset{\sim}{\Gamma}_a \underset{\sim}{\chi}$ is also a matrix of antiexact forms. The homotopy operator H inverts the operator d on the module of antiexact forms [see(2.2.5)], so that (3.3.1) implies

$$\underset{\sim}{\Gamma}_a \underset{\sim}{\chi} = H(d\underset{\sim}{\Gamma}_a \underset{\sim}{\chi}) - H(\underset{\sim}{\Gamma}_a \wedge d\underset{\sim}{\chi}) ,$$

that is

(3.3.2) $H(\underset{\sim}{\Gamma}_a \wedge d\underset{\sim}{\chi}) = H(d\underset{\sim}{\Gamma}_a \underset{\sim}{\chi}) - \underset{\sim}{\Gamma}_a \underset{\sim}{\chi}$.

Thus, with (3.3.2) and the fact that we have agreed to put $\underset{\sim}{A} = \underset{\sim}{I}$, the system (3.1.4-7) can be written in the following way:

(3.3.3) $\underset{\sim}{B} = d\underset{\sim}{\chi} + \underset{\sim}{\Gamma}_a \underset{\sim}{\chi} + \underset{\sim}{\eta} - H(d\underset{\sim}{\Gamma}_a \underset{\sim}{\chi})$,

(3.3.4) $\underset{\sim}{\mathcal{D}} = d\underset{\sim}{\eta} + H(d\underset{\sim}{\Gamma}_a \wedge d\underset{\sim}{\chi}) + \underset{\sim}{\Gamma}_a \wedge [\underset{\sim}{\eta} - H(d\underset{\sim}{\Gamma}_a \underset{\sim}{\chi}) + \underset{\sim}{\Gamma}_a \underset{\sim}{\chi}]$,

(3.3.5) $\underset{\sim}{\mathcal{K}} = \underset{\sim}{\Gamma}_a \wedge [d\underset{\sim}{\chi} + \underset{\sim}{\Gamma}_a \underset{\sim}{\chi} + \underset{\sim}{\eta} - H(d\underset{\sim}{\Gamma}_a \underset{\sim}{\chi})]$,

$$(3.3.6)\quad \underset{\sim}{\Omega} = (d\underset{\sim}{\Gamma}_a + \underset{\sim}{\Gamma}_a \wedge \underset{\sim}{\Gamma}_a) \wedge [d\underset{\sim}{\chi} + \underset{\sim}{\Gamma}_a \underset{\sim}{\chi} + \underset{\sim}{\eta} - H(d\underset{\sim}{\Gamma}_a \underset{\sim}{\chi})]$$

$$- \underset{\sim}{\Gamma}_a \wedge \{d\underset{\sim}{\eta} + H(d\underset{\sim}{\Gamma}_a \wedge d\underset{\sim}{\chi}) + \underset{\sim}{\Gamma}_a \wedge [\underset{\sim}{\eta} + \underset{\sim}{\Gamma}_a \underset{\sim}{\chi} - H(d\underset{\sim}{\Gamma}_a \underset{\sim}{\chi})]\} .$$

Notice that the matrix $\underset{\sim}{\Gamma}_a$ of antiexact 1-forms occurs in $\underset{\sim}{B}$ only in conjunction with the state vector $\underset{\sim}{\chi}$ and acts on $\underset{\sim}{\chi}$ from the left (recall that the subscript "a" stands for "antiexact part", it is not a summation index). If we look at just the first two terms on the right-hand side of (3.3.3) it is reminicent of the expression for the absolute time derivative $\frac{d}{dt} + \vec{\omega}\times$ in a rotating frame of reference. Thus, it is natural to expect that $\underset{\sim}{\Gamma}_a$ is related to the internal rotation degrees of freedom. This expectation will be shown to be correct in the next section. Some such relation is clearly necessary, for the quantities $\underset{\sim}{\Gamma}_a$ are, at this point, arbitrary; they have not as yet been associated with any specific physical aspect of defect dynamics.

Before turning to the determination of the quantities $\underset{\sim}{\Gamma}_a$, we need to justify the assumption that $\underset{\sim}{\Gamma}_a$ is a matrix of *antiexact* 1-forms. We first note that the representation (3.3.3-6) of the solution of the Cartan structure equations involves $\underset{\sim}{\Gamma}_a$, while the $\underset{\sim}{\Gamma}$ that occurs in the Cartan structure equations (2.3.1) is a matrix of 1-forms that is not necessarily antiexact. However, once we have arrived at the representation (3.3.3), it follows that $d\chi$ is the exact and hence the completely integrable part of $\underset{\sim}{B}$, while the remaining terms, $\Gamma_a\chi + \eta - H(d\Gamma_a\chi)$, constitute the antiexact part of $\underset{\sim}{B}$. Now, this decomposition of $\underset{\sim}{B}$ into exact and antiexact parts is unique, [see 2.2.5)], while it has been shown [5] that the additive 45-parameter gauge group of defect dynamics can always be used to achieve this explicit decomposition of the various terms that comprise the fields $\underset{\sim}{B}$, $\underset{\sim}{D}$, $\underset{\sim}{K}$ and $\underset{\sim}{\Omega}$. Further, the attitude matrix $\underset{\sim}{A}$ associated with $\underset{\sim}{\Gamma}$ satisfies the Riemann-Graves integral equation $\underset{\sim}{A} = \underset{\sim}{I} - H(\underset{\sim}{\Gamma}A)$ [see(2.5.6)]. With $\underset{\sim}{\Gamma} = \underset{\sim}{\Gamma}_a$ antiexact, $\underset{\sim}{\Gamma}_a \underset{\sim}{A}$ is also antiexact and hence belongs to ker H . Thus, $\underset{\sim}{A}$ is given by $\underset{\sim}{A} = \underset{\sim}{I}$ as previously required. Indeed $\underset{\sim}{\Gamma} = \underset{\sim}{\Gamma}_a$ is necessary in order that $\underset{\sim}{A} = \underset{\sim}{I}$.

Having made sufficient note of this fact, we will omit the subscript "a" on the connection matrix of antiexact 1-forms. Whenever $\underset{\sim}{\Gamma}$ occurs in the sequel it is to be understood as an element of the module $A^1_{3,3}(E_4)$ of matrices of antiexact 1-forms.

3.4 The Minimal Replacement Argument for SO(3)

We gave a brief summary in Section 2.3 of the Yang-Mills minimal replacement that was originally formulated for a simi-simple Lie group of internal symmetries. On the otherhand, we have shown in Section 2.6 that the underlying group of elasticity theory is $G_o = SO(3)_o \triangleright T(3)_o$. Since $SO(3)_o$ is a subgroup of G_o that is semi-simple, it is possible to apply the Yang-Mills minimal replacement argument directly for the subgroup $SO(3)_o$. The reader should note that there is no real choice here if the homogeneity of the action of $SO(3)_o$ is to be broken. If the minimal replacement is not made, the Lagrangian of elasticity theory will cease to be invariant under the action of the transformation group G. However, it is exactly the invariance of the Lagrangian under the action of G that gives rise to the fact that the resulting field equations imply laws of balance for linear momentum and moment of momentum via the Noether theorem. Accordingly, it is only through the minimal replacement construct that the field equations will continue to imply laws of balance for linear momentum and moment of momentum when the homogeneity of the action of the group G_o is broken. In other words, breaking of the homogeneity of the action of G_ρ without the minimal replacement construct will lead to field equations that do not describe the fundamental properties of mechanics.

Let $\underset{\sim}{\gamma}_\alpha$ denote constant-valued matrices that form a basis for the Lie algebra of the 3-by-3 matrix representation of $SO(3)_0$. Upon breaking the homogeneity of the action of $SO(3)_0$ (i.e., we replace $'\underset{\sim}{\chi}(X^a) = \underset{\sim}{A}\underset{\sim}{\chi}(X^a)$ by $'\underset{\sim}{\chi}(X^b) = \underset{\sim}{A}(X^b)\underset{\sim}{\chi}(X^b))$ we have to introduce the Yang Mills compensating fields of matrices $\underset{\sim}{\Gamma}_y$ of 1-forms that take values in the Lie algebra of $SO(3)$. This implies that we may express $\underset{\sim}{\Gamma}_y$ in the form

$$(3.4.1) \qquad \underset{\sim}{\Gamma}_y = W^\alpha \underset{\sim}{\gamma}_\alpha , \quad W^\alpha \varepsilon \Lambda^1(E_4) .$$

Accordingly, (2.3.7), (2.3.10) and (2.3.26) show that the minimal replacement associated with the breaking of the homogeneity of the action of $SO(3)_0$ is given by

$$(3.4.2) \qquad d\underset{\sim}{\chi} \rightarrow D\underset{\sim}{\chi} = d\underset{\sim}{\chi} + \underset{\sim}{\Gamma}_y \underset{\sim}{\chi} ,$$

since a matrix representation of $SO(3)$ acts on the state

vector $\underset{\sim}{\chi}$ from the left (see(2.3.27)).

In defect dynamics, the deformation gradient matrix, $d\underset{\sim}{\chi}$, is replaced by the distortion, $\underset{\sim}{B}$. By (3.4.2), the minimal replacement for SO(3) gives

$$(3.4.3) \qquad \underset{\sim}{B} = D\underset{\sim}{\chi} = d\underset{\sim}{\chi} + \underset{\sim}{\Gamma}_y \underset{\sim}{\chi} .$$

A comparison of the representation (3.3.3) with (3.4.3) shows that we achieved the explicit identification

$$(3.4.4) \qquad \underset{\sim}{\Gamma} = \underset{\sim}{\Gamma}_y = W^\alpha \underset{\sim}{\gamma}_\alpha$$

and that $\underset{\sim}{\eta}$ must therefore satisfy

$$(3.4.5) \qquad \underset{\sim}{\eta} = H(d\underset{\sim}{\Gamma}\ \underset{\sim}{\chi}) = H((d\underset{\sim}{\Gamma} + \underset{\sim}{\Gamma}\wedge\underset{\sim}{\Gamma})\underset{\sim}{\chi}) = H(\underset{\sim}{\Theta}\underset{\sim}{\chi}) .$$

The components of the matrix $\underset{\sim}{\Gamma}$ of antiexact 1-forms can be written in the following way

$$(3.4.6) \qquad \Gamma^i_j = W^\alpha \gamma^i_{\alpha j} = W^\alpha_b\ \gamma^i_{\alpha j}\ dX^b = \Gamma^i_{bj}\ dX^b , \quad W^\alpha \varepsilon A^1(E_4) .$$

Hence, the 27 unknown function Γ^i_{bj} (recall that there are 36 unknowns Γ^i_{bj}, but they have to satisfy 9 algebraic conditions $X^b\Gamma^i_{bj} = 0$) are reduced to 12 variables W^α_a. However, the three 1-forms W^α are antiexact so that the functions W^α_b have to satisfy the three algebraic conditions (for $X^a_o = 0$)

$$(3.4.7) \qquad X^b\ W^\alpha_b = 0 ,$$

which in turn reduces the number of unknowns to 9.

Now by (3.4.5) and (2.5.1) the system (3.3.3-6) becomes

$$\begin{aligned} \underset{\sim}{B} &= d\underset{\sim}{\chi} + \underset{\sim}{\Gamma}\underset{\sim}{\chi} = \underset{\sim}{D}\underset{\sim}{\chi} \\ \underset{\sim}{\mathcal{D}} &= (d\underset{\sim}{\Gamma} + \underset{\sim}{\Gamma} \wedge \underset{\sim}{\Gamma})\underset{\sim}{\chi} = \underset{\sim}{\Theta}\underset{\sim}{\chi} \\ \underset{\sim}{K} &= \underset{\sim}{\Gamma}\wedge(d\underset{\sim}{\chi} + \underset{\sim}{\Gamma}\underset{\sim}{\chi}) = \underset{\sim}{\Gamma}\wedge\underset{\sim}{D}\underset{\sim}{\chi} \\ \underset{\sim}{\Omega} &= d(\underset{\sim}{\Gamma}\wedge D\underset{\sim}{\chi}) \ . \end{aligned} \tag{3.4.8}$$

(recall that $\underset{\sim}{\Theta}$ is the matrix of curvature 2-forms associated with $\underset{\sim}{\Gamma}$) .

The state described by equations (3.4.8) is a state generated by disclinations since the connection matrix $\underset{\sim}{\Gamma}$ accounts for the breaking of the homogeneity of the action of the rotation group. This is also evident when we write out the first of (3.4.8) so as to obtain

$$\underset{\sim}{B} = d\underset{\sim}{\chi} + W^{\alpha}\underset{\sim}{\gamma}_{\alpha}\underset{\sim}{\chi} \ ;$$

for $d\underset{\sim}{\chi}$ describes the integrable part of the distortion, while $W^{\alpha}\underset{\sim}{\gamma}_{\alpha}\underset{\sim}{\chi}$ describes the nonintegrable local rotation (internal spin) acting on the instantaneous state vector.

The second of the equations (3.4.8) shows that the state characterized by the inhomogeneous action of SO(3) is not a state of pure disclination since $\underset{\sim}{\mathcal{D}} \neq \underset{\sim}{0}$. Recall that the matrix $\underset{\sim}{\mathcal{D}}$ characterizes dislocations, as follows directly from (2.7.8)

$$\mathcal{D}^i = J^i \wedge dT + \alpha^i ,$$

where J^i are the dislocation current 1-forms and α^i are the dislocation density 2-forms. In fact, it follows from (3.4.8) that disclinations always generate dislocations. For the purposes of this discussion, it is useful to refer to dislocations generated by disclinations as *disclination driven dislocations*. From these elementary considerations, we conclude that it is not possible to have a pure disclinated material. Whenever disclinations are present, dislocations are present of necessity. However, as we shall see later, it is possible to have a pure dislocated medium. In fact, it is this state, the pure dislocated state, that is usually treated in the literature.

The reader should note that the disclination driven dislocations have the particularly simple form

$$\underset{\sim}{\mathcal{D}} = \underset{\sim}{\Theta}\underset{\sim}{\chi} ;$$

that is, the dislocation 2-forms are realized by the curvature of the SO(3) connection acting as a linear operator on the state vector $\underset{\sim}{\chi}$. Since $\underset{\sim}{\Theta} = d\underset{\sim}{\Gamma} + \underset{\sim}{\Gamma} \wedge \underset{\sim}{\Gamma}$ and $\underset{\sim}{\Gamma}$ is antiexact, we have

$$\underset{\sim}{\mathcal{D}} = (d\underset{\sim}{\Gamma})\underset{\sim}{\chi} + \underset{\sim}{\Gamma}\wedge\underset{\sim}{\Gamma}\underset{\sim}{\chi}$$

and the terms $\underset{\sim}{\Gamma}\wedge\underset{\sim}{\Gamma}\underset{\sim}{\chi}$ are antiexact in view of the module property of antiexact forms. It thus follows that $\underset{\sim}{\Gamma}\wedge\underset{\sim}{\Gamma}\underset{\sim}{\chi}$ belongs to ker H and hence (see(3.4.5))

$$H\underset{\sim}{\mathcal{D}} = H(d\underset{\sim}{\Gamma}\ \underset{\sim}{\chi}) = \underset{\sim}{\eta}$$

In this regard, we also note that $\underset{\sim}{\Theta} = F^{\alpha}\ \underset{\sim}{\gamma}_{\alpha}$, $F^{\alpha} = dW^{\alpha} + C^{\alpha}_{\beta\rho}\ W^{\beta}\wedge W^{\rho}/2$ yield

$$\underset{\sim}{\mathcal{D}} = F^{\alpha}\ \underset{\sim}{\gamma}_{\alpha}\underset{\sim}{\chi}\ ,$$

and hence the disclination driven dislocation 2-forms may be viewed as linear combinations of the three effective state vectors $\underset{\sim}{\gamma}_{\alpha}\underset{\sim}{\chi}$ with the 2-form valued coefficients F^{α} .

We stress again the importance of the requirement that the entries W^{α} of the matrix $\underset{\sim}{\Gamma} = W^{\alpha}\underset{\sim}{\gamma}_{\alpha}$ be antiexact. If this were not the case, we could use (2.2.5b) to write

$$\underset{\sim}{B} = d[\underset{\sim}{\chi} + H(\underset{\sim}{\Gamma}\underset{\sim}{\chi})] + Hd(\underset{\sim}{\Gamma}\underset{\sim}{\chi})\ ,$$

so that the matrix $\underset{\sim}{\chi}$ would not be the total integrable displacement and all of the identifications would dissolve in confusion. The careful reader will have noted in this matter that the Yang-Mills theory does not make the assumption that the compensating Yang-Mills 1-forms W^{α} , are antiexact. However, there is no real difficulty here, for it has been shown in section 2.4 that there always exists a choice of gauge for which any system of compensating Yang-Mills 1-forms becomes a system of antiexact 1-forms.

3.5 The Minimal Replacement Argument for SO(3)⊳T(3)

Up to this point, the discussion was based on the minimal coupling theory as stated in Section 2.3; namely, the gauge group was taken to be the semi-simple group SO(3) and the connection matrix $\underset{\sim}{\Gamma}$ occurs due to the breaking of the homogeneity of the action of this rotation group. We know, however, that defect dynamics is an outgrowth of elasticity theory that has the underlying group $G_o = SO(3)_o \triangleright T(3)_o$ that is a semi-direct product of the rotation group $SO(3)_o$ and the translation group $T(3)_o$. The group G_o is not semi-simple and it does not have a faithful matrix representation on the space of integrable displacements $\underset{\sim}{\chi}$. Thus, we require a nontrivial extension of the Yang-Mills theory. Also, it is important to note that the action of $SO(3)_o$ is quite different from the action of $T(3)_o$; elements of $SO(3)_o$ act multiplicatively on $\underset{\sim}{\chi}$ from the left while elements of $T(3)_o$ act additively:

$$'\underset{\sim}{\chi} = A\underset{\sim}{\chi} + \underset{\sim}{b} \ , \quad \underset{\sim}{A} \varepsilon SO(3)_o \ , \quad \underset{\sim}{b} \varepsilon T(3)_o \, .$$

In the previous section, the minimal replacement (3.4.2) takes account only of the breaking of the homogeneity of the action of $SO(3)_o$, and therefore says nothing about the breaking of the homogeneity of the action of the translation group $T(3)_o$. However, it is clear, physically, that dislocations are translational defects and may be viewed as responses to the breaking of the homogeneity of the action of the translation group. Therefore, it is natural to amend the minimal replacement (3.4.2) for just SO(3) so that it reads

$$(3.5.1) \quad d\underset{\sim}{\chi} \longrightarrow d\underset{\sim}{\chi} + \underset{\sim}{\Gamma}\underset{\sim}{\chi} + \underset{\sim}{\phi} \,.$$

Here, $\underset{\sim}{\phi} \in \Lambda_{3,1}(E_4)$ is a column matrix of 1-forms that compensates for the breaking of the homogeneity of the action of $T(3)_o$ and hence describes dislocations. This, then leads to the final expression for the distortion

$$(3.5.2) \quad \underset{\sim}{B} = D\underset{\sim}{\chi} + \underset{\sim}{\phi} = d\underset{\sim}{\chi} + \underset{\sim}{\Gamma}\underset{\sim}{\chi} + \underset{\sim}{\phi} \,.$$

The distortion $\underset{\sim}{B}$ thus arises from three sources: (1) from the totally integrable response, $d\underset{\sim}{\chi}$; (2) from the inhomogeneous action of the rotation group, $\underset{\sim}{\Gamma}\underset{\sim}{\chi} = W^{\alpha}\underset{\sim}{\gamma}_{\alpha}\underset{\sim}{\chi}$; and $(3)_o$ from the breaking of the homogeneity of the action of the translation group, $\underset{\sim}{\phi}$. A comparison of (3.5.2) with (3.3.3) shows that the previous expression (3.4.5) for $\underset{\sim}{\eta}$ must now be replaced by

$$(3.5.3) \quad \underset{\sim}{\eta} = H(d\underset{\sim}{\Gamma}\underset{\sim}{\chi}) + \underset{\sim}{\phi} = H((d\underset{\sim}{\Gamma} + \underset{\sim}{\Gamma}\wedge\underset{\sim}{\Gamma})\underset{\sim}{\chi} + d\underset{\sim}{\phi} + \underset{\sim}{\Gamma}\wedge\underset{\sim}{\phi})$$

We note that of necessity the vector $\underset{\sim}{\phi}$ has to be an element of the submodule $A^1_{3,1}(E_4)$ of antiexact 1-forms. If this were not the case we could write [see(2.2.5b)]

$$\underset{\sim}{\phi} = dH\underset{\sim}{\phi} + Hd\underset{\sim}{\phi}$$

so that $\underset{\sim}{\chi}$ would no longer represent the total integrable displacement. There is no real difficulty here, for there is always an additive gauge by which any given $\underset{\sim}{\phi}$ may be replaced by an antiexact one [5].

In the absence of disclinations the minimal replacement (3.5.1) reduces to

$$(3.5.4) \qquad d\underset{\sim}{\chi} \longrightarrow d\underset{\sim}{\chi} + \underset{\sim}{\phi} = \underset{\sim}{B} \ .$$

From (3.4.19) and (3.1.1) we obtain

$$(3.5.5) \qquad \underset{\sim}{\Sigma} = \underset{\sim}{\mathcal{D}} = d\underset{\sim}{\phi} = d\underset{\sim}{B} \ .$$

In this case $\underset{\sim}{\phi}$ uniquely determines the Cartan torsion $\underset{\sim}{\Sigma}$. Our results are thus seen to be consistent with the description of dislocations by means of Cartan torsion that is reported in the literature [10,18,31].

The expression (3.5.2) for the distortion is to be the starting point for the theory developed in this work. It provides a complete description of both dislocations and disclinations. In view of the importance of this result, it would be useful to have an independent derivation of (3.5.2). The next section gives just that; namely, a direct derivation of the minimal replacement for the group $G = SO(3) \triangleright T(3)$ from a group-theoretic point of view.

3.6 An Independent Group-Theoretic Argument

Let us for the moment go back to Section 2.3. In order to introduce the operators D_b and the compensating fields $\underset{\sim}{\Gamma}_b$ that transform under the inhomogeneous action of a group G through

$$'(D_b\underset{\sim}{\psi}) = \underset{\sim}{A}(D_b\underset{\sim}{\psi}) \ ,$$

$$'\underset{\sim}{\Gamma}_b = \underset{\sim}{A}\underset{\sim}{\Gamma}_b\underset{\sim}{A}^{-1} - (\partial_b\underset{\sim}{A})\underset{\sim}{A}^{-1} \ , \quad \underset{\sim}{A}\varepsilon G \ ,$$

it was assumed that G had a faithful matrix representation as action from the left on the state vector $\underset{\sim}{\psi}$. It was thus simple to proceed in Section 3.4 with the "minimal replacement" construct since the underlying group was the semi-simple rotation group SO(3) that acted from the left as a matrix group on the state vector $\underset{\sim}{\chi}$. Now, the discussion given in Section 2.3 shows that it is not semi-simplicity of the group that is important, but rather the existence of a faithful matrix representation as action from the left on the state vector that plays the essential role.

The full group $G_o = SO(3)_o \triangleright T(3)_o$ of elasticity, and hence of defect dynamics, does not have a faithful matrix representation in the space of state vectors $\underset{\sim}{\chi}$. Thus, we need to find a new space of state vectors such that G_o

will admit a faithful matrix representation by action on the left. Once we have done this, it is then a simple matter to proceed with the "minimal replacement" construct.

To this end, we consider the vector space V_4 and the affine subset V of V_4 consisting of all vectors of the form

$$(3.6.1) \qquad \hat{\underset{\sim}{\chi}} = \begin{Bmatrix} \underset{\sim}{\chi} \\ 1 \end{Bmatrix} = \begin{Bmatrix} \chi^1 \\ \chi^2 \\ \chi^3 \\ 1 \end{Bmatrix} .$$

It is clear, from (3.6.1), that each state vector $\underset{\sim}{\chi}$ in V_3 determines a unique $\hat{\underset{\sim}{\chi}}$ in $V \subset V_4$, and conversly. We note, however, that the set V is not a vector subspace of V_4 since it is not closed under vector space operations. Nevertheless, the action of 4-by-4 matrices on elements of V is well defined since each element of V belongs to V_4 .

Let M denote the collection of all 4-by-4 matrices of the form [14]

$$(3.6.2) \qquad \underset{\sim}{M} = \begin{pmatrix} \underset{\sim}{A} & \{\underset{\sim}{b}\} \\ [\underset{\sim}{0}] & 1 \end{pmatrix} .$$

Here, $\underset{\sim}{A}$ is an orthogonal 3-by-3 matrix, $\{\underset{\sim}{b}\}$ is a column matrix of dimension 3, and $[\underset{\sim}{0}]$ is a zero row matrix of dimension 3. Since

$$(3.6.3) \qquad \begin{pmatrix} \underset{\sim}{A} & \{\underset{\sim}{b}\} \\ [\underset{\sim}{0}] & 1 \end{pmatrix} \begin{Bmatrix} \underset{\sim}{\chi} \\ 1 \end{Bmatrix} = \begin{Bmatrix} \underset{\sim}{A}\underset{\sim}{\chi} + \underset{\sim}{b} \\ 1 \end{Bmatrix} = \begin{Bmatrix} {}'\underset{\sim}{\chi} \\ 1 \end{Bmatrix}$$

any element of M mapps V into V and realizes the action of $SO(3) \rhd T(3)$ on $\underset{\sim}{\chi}$ through its action on the corresponding $\hat{\underset{\sim}{\chi}} \varepsilon V$. If $\underset{\sim}{M}_1$ and $\underset{\sim}{M}_2$ belong to M , with

$$(3.6.4) \qquad \underset{\sim}{M}_1 = \begin{pmatrix} \underset{\sim}{A}_1 & \{\underset{\sim}{b}_1\} \\ [\underset{\sim}{0}] & 1 \end{pmatrix} , \quad M_2 = \begin{pmatrix} \underset{\sim}{A}_2 & \{\underset{\sim}{b}_2\} \\ [\underset{\sim}{0}] & 1 \end{pmatrix}$$

then

$$(3.6.5) \qquad \underset{\sim}{M}_1 \underset{\sim}{M}_2 = \begin{pmatrix} \underset{\sim}{A}_1 \underset{\sim}{A}_2 & \underset{\sim}{A}_1 \{\underset{\sim}{b}_2\} + \{\underset{\sim}{b}_1\} \\ [\underset{\sim}{0}] & 1 \end{pmatrix} \varepsilon M .$$

Thus, the collection M forms a matrix Lie group, and (3.6.3) shows that this matrix group forms a faithful matrix representation of $SO(3) \rhd T(3)$ by left action on the elements of the affine set V . This achieves the desired result, and we shall now simply write the matrix group G for the set M .

The infinitesimal generators of the matrix group G are

$$(3.6.6) \qquad \begin{pmatrix} \underset{\sim}{\gamma}_\alpha & \{\underset{\sim}{0}\} \\ [\underset{\sim}{0}] & 0 \end{pmatrix}, \quad \alpha = 1,2,3 \; ; \quad \begin{pmatrix} \underset{\sim}{0} & \{\underset{\sim}{t}_i\} \\ [\underset{\sim}{0}] & 0 \end{pmatrix}, \quad i = 1,2,3 \, .$$

Here, $\underset{\sim}{\gamma}_\alpha$ are the 3-by-3 infinitesimal generating matrices of the rotation group SO(3) and $\underset{\sim}{t}_i = [\delta_{i1}, \delta_{i2}, \delta_{i3}]^T$ are the infinitesimal generating matrices of the translation group T(3) .

Now that we have a faithful matrix representation of G , we can proceed exactly as in Section 2.3. The covariant exterior derivative in the larger space V_4 is defined for all elements of V by

$$(3.6.7) \qquad \hat{D}\hat{\underset{\sim}{X}} = d\hat{\underset{\sim}{X}} + \hat{\underset{\sim}{\Gamma}}\hat{\underset{\sim}{X}} \, ,$$

where $\hat{\underset{\sim}{\Gamma}}$ is a 4-by-4 matrix of connection 1-forms that take values in the Lie algebra of G . Thus, $\hat{\underset{\sim}{\Gamma}}$ is given by

$$(3.6.8) \qquad \hat{\underset{\sim}{\Gamma}} = W^\alpha \begin{pmatrix} \underset{\sim}{\gamma}_\alpha & \{\underset{\sim}{0}\} \\ [\underset{\sim}{0}] & 0 \end{pmatrix} + \phi^i \begin{pmatrix} \underset{\sim}{0} & \{\underset{\sim}{t}_i\} \\ [\underset{\sim}{0}] & 0 \end{pmatrix} .$$

Here, W^α and ϕ^i are 1-forms on the space E_4 . When (3.6.1) and (3.6.8) are used to express the various terms that occur in (3.6.7), we obtain

$$(3.6.9) \qquad \hat{D}\hat{\underset{\sim}{X}} = \begin{Bmatrix} d\underset{\sim}{X} + W^\alpha \underset{\sim}{\gamma}_\alpha \underset{\sim}{X} + \phi^i \underset{\sim}{t}_i \\ 0 \end{Bmatrix} = \begin{Bmatrix} D\underset{\sim}{X} + \phi^i \underset{\sim}{t}_i \\ 0 \end{Bmatrix} .$$

The 1-form valued column matrix

$$(3.6.10) \qquad \underset{\sim}{\phi} = \phi^i \underset{\sim}{t}_i$$

is thus naturally introduced. It is also clear from (3.6.9) that the "minimal replacement" construct now takes the form

$$(3.6.11) \qquad d\underset{\sim}{\chi} \longrightarrow D\underset{\sim}{\chi} + \underset{\sim}{\phi} = d\underset{\sim}{\chi} + \underset{\sim}{\Gamma}\underset{\sim}{\chi} + \underset{\sim}{\phi}$$

which agrees exactly with what was obtained in the previous section, but now with the added confidence of an independent group theoretic derivation.

We know that the matrix $\hat{\underset{\sim}{\Gamma}}$ transforms under the action of an element of G with associated matrix $\underset{\sim}{M}$ according to [see(2.3.9)]

$$(3.6.12) \qquad '\hat{\underset{\sim}{\Gamma}} = \underset{\sim}{M}\hat{\underset{\sim}{\Gamma}}\underset{\sim}{M}^{-1} - d\underset{\sim}{M}\,\underset{\sim}{M}^{-1} .$$

Now, (3.6.8) and (3.6.12) give

$$'\hat{\underset{\sim}{\Gamma}} = \begin{pmatrix} '\underset{\sim}{\Gamma} & '\underset{\sim}{\phi} \\ [\underset{\sim}{0}] & 0 \end{pmatrix} = \left[\begin{pmatrix} \underset{\sim}{A} & \underset{\sim}{b} \\ [\underset{\sim}{0}] & 1 \end{pmatrix}\begin{pmatrix} \underset{\sim}{\Gamma} & \underset{\sim}{\phi} \\ [\underset{\sim}{0}] & 0 \end{pmatrix}\right.$$

$$\left. - \begin{pmatrix} d\underset{\sim}{A} & d\underset{\sim}{b} \\ [\underset{\sim}{0}] & 0 \end{pmatrix}\right]\begin{pmatrix} A^{-1} & -A^{-1}b \\ [\underset{\sim}{0}] & 1 \end{pmatrix}$$

$$= \begin{pmatrix} \underset{\sim}{A}\underset{\sim}{\Gamma}\underset{\sim}{A}^{-1} - d\underset{\sim}{A}\underset{\sim}{A}^{-1} & \underset{\sim}{A}\underset{\sim}{\phi} - d\underset{\sim}{b} - (\underset{\sim}{A}\underset{\sim}{\Gamma} - d\underset{\sim}{A})\underset{\sim}{A}^{-1}\,\underset{\sim}{b} \\ [\underset{\sim}{0}] & 0 \end{pmatrix} .$$

We thus obtain the already known result

$$(3.6.13) \qquad '\underset{\sim}{\Gamma} = \underset{\sim}{A}\underset{\sim}{\Gamma}\underset{\sim}{A}^{-1} - d\underset{\sim}{A}\,\underset{\sim}{A}^{-1} ,$$

and in addition, the transformation law for the vector $\underset{\sim}{\phi}$:

$$(3.6.14) \qquad '\underset{\sim}{\phi} = \underset{\sim}{A}\underset{\sim}{\phi} - d\underset{\sim}{b} - (\underset{\sim}{A}\underset{\sim}{\Gamma}\underset{\sim}{A}^{-1} - d\underset{\sim}{A}\,\underset{\sim}{A}^{-1})\underset{\sim}{b} .$$

It is then evident from (3.6.3), (3.6.13) and (3.6.14) that the distortion 1-forms $\underset{\sim}{B}$ transform according to

$$(3.6.15) \qquad '\underset{\sim}{B} = d'\underset{\sim}{\chi} + '\underset{\sim}{\phi} + '\underset{\sim}{\Gamma}\,'\underset{\sim}{\chi} = \underset{\sim}{A}\underset{\sim}{B} ,$$

so that it is covariant under the action of the group G .

The group-theoretic derivation just given has allowed us to introduce a complete characterization of the breaking of the homogeneity of the action of both the rotation group $SO(3)_0$ and the translation group $T(3)_0$. Since the distortion arises through such breakings of the homogeneity of action of the underlying group G_0, the "minimal replacement" argument gives us the following specific replacement for the displacement gradient of elasticity theory:

$$(3.6.16) \qquad d\underset{\sim}{\chi} \longrightarrow \underset{\sim}{B} = D\underset{\sim}{\chi} + \underset{\sim}{\phi} = d\underset{\sim}{\chi} + \underset{\sim}{\Gamma}\underset{\sim}{\chi} + \underset{\sim}{\phi} .$$

In particular, we see that the breaking of the homogeneity of the multiplicative action of the group G on $\hat{\underset{\sim}{\chi}}$ from the left via the connection matrix $\hat{\underline{\Gamma}}$ splits naturally into the multiplicative action of the group SO(3) on the state vector $\underline{\chi}$ from the left via the connection $\underline{\Gamma}$, and an additive action of the group T(3) via the 1-forms $\underline{\phi}$. Thus, the term $\underline{\Gamma}\underline{\chi}$ represents the non-integrable internal rotation degrees of freedom, namely disclinations, while the term $\underline{\phi}$ represents the non-integrable internal translation degrees of freedom, namely dislocations. We thus have come full circle with complete agreement with the more physically based agrument given in previous sections.

The view from the affine set V of V_4 has shown that $\hat{D}\hat{\underline{\chi}}$ and $\hat{\underline{\Gamma}}$ transform under the action of G by

$$'\hat{D}'\hat{\underline{\chi}} = \underline{M}D\hat{\underline{\chi}} \; , \quad '\hat{\underline{\Gamma}} = (\underline{M}\hat{\underline{\Gamma}} - d\underline{M})\underline{M}^{-1} \; , \quad \underline{M} \; \varepsilon \; G$$

while $\hat{D}\hat{\chi} = \{\underline{B}^T, \; 0\}^T$ and $\hat{\underline{\Gamma}}$ takes its values in the Lie algebra of G . Accordingly, the results established at the end of Section 2.5 show that the Cartan structure equations generated from $\hat{D}\chi$ and $\hat{\underline{\Gamma}}$ are gauge covariant. Thus, we may impose the antiexact gauge by appropriate gauge transformations, in which case (3.6.8) shows that *the 1-forms* W^α *and* ϕ^i *may be taken to be antiexact* without loss of generality. Further, and of more importance, it was shown at the end of Section 2.4 that a change of center for the construction of the linear homotopy operator could be achieved by an appropriate gauge transformation. Gauge covariance of the Cartan structure equations formed from $\hat{D}\hat{\underset{\sim}{\chi}}$ and $\hat{\Gamma}$ thus allows us to pick any convenient center for E_4 without substantial change in gauge covariant quantities such as $\hat{D}\hat{\underset{\sim}{\chi}}$, $\hat{\underset{\sim}{\Sigma}}$, and $\hat{\underset{\sim}{\Theta}}$. Once the new center is in place, further gauge transformations may be used to enforce the antiexact gauge relative to the new linear homotopy operator associated with new center: *any center of* E_4 *can be used for the construction of the gauge theory of defects.* This does not mean that values of various field quantities at specific points will not change for different choices of centers, for they clearly must change in accordance with how the gauge group acts on them. Rather, such changes are inconsequential since the theory and the field equations of the theory will turn out to be gauge covariant.

3.7 Field Variables and Observables

With the minimal replacement (3.5.1) and the choice (3.5.3) for $\underset{\sim}{\eta}$ we rewrite the equations (3.4.8) in terms of the 27 unknown functions χ^i , W^α_a , ϕ^i_a :

$$\underset{\sim}{B} = d\underset{\sim}{\chi} + \underset{\sim}{\Gamma}\underset{\sim}{\chi} + \underset{\sim}{\phi} = D\underset{\sim}{\chi} + \underset{\sim}{\phi} ,$$

$$\underset{\sim}{\mathcal{D}} = \underset{\sim}{\Theta}\underset{\sim}{\chi} + D\underset{\sim}{\phi} , \tag{3.7.1}$$

$$\underset{\sim}{K} = \underset{\sim}{\Gamma}\wedge(D\underset{\sim}{\chi} + \underset{\sim}{\phi}) ,$$

$$\underset{\sim}{\Omega} = d(\underset{\sim}{\Theta}\underset{\sim}{\chi} + D\underset{\sim}{\phi}) .$$

The functions W^α_a and ϕ^i_a are subject to the algebraic conditions

$$\chi^a \, W^\alpha_a = 0 , \quad \chi^a \, \phi^i_a = 0 , \tag{3.7.2}$$

since the 1-forms W^α and ϕ^i are required to be anti-exact. Equations (3.7.1) can also be written in terms of the components of the matrices involved as follows

$$B^i = d\chi^i + \Gamma^i_j \, \chi^j + \phi^i = d\chi^i + W^\alpha \gamma^i_{\alpha j} \, \chi^j + \phi^i$$

$$\mathcal{D}^i = \Theta^i_j \chi^j + D\phi^i = F^\alpha \gamma^i_{\alpha j} \, \chi^j + D\phi^i$$

$$= (dW^\alpha + \tfrac{1}{2} \, C^\alpha_{\beta\gamma} W^\beta \wedge W^\gamma)\gamma^i_{\alpha j} \, \chi^j + D\phi^i \tag{3.7.3}$$

$$K^i = \Gamma^i_j \wedge (D\chi^j + \phi^j) = W^\alpha \gamma^i_{\alpha j} \wedge (D\chi^j + \phi^j)$$

$$\Omega^i = d\mathcal{D}^i .$$

Notice the inseparability of the "rotation" and "translation" dislocations. Namely, in the expression

$$\underset{\sim}{B} = d\underset{\sim}{\chi} + (\underset{\sim}{\Gamma}\underset{\sim}{\chi} + \underset{\sim}{\phi})$$

for the distortion, the vector $\underset{\sim}{\phi}$ characterizes the translation dislocations, while the presence of disclinations, characterized by $\underset{\sim}{\Gamma}\underset{\sim}{\chi}$, generates the so called rotation or disclination driven dislocations (see Sec. 3.4). The entries of the square matrix $\underset{\sim}{\Gamma}$ and the column vector $\underset{\sim}{\phi}$ are antiexact 1-forms, and so is the sum $\underset{\sim}{\Gamma}\underset{\sim}{\chi} + \underset{\sim}{\phi}$. By (2.2.5), (2.5.1) and (3.1.1) we obtain

$$(3.7.4) \qquad \underset{\sim}{\Gamma}\underset{\sim}{\chi} + \underset{\sim}{\phi} = \underset{\sim}{B}_a = Hd\underset{\sim}{B} = H(\underset{\sim}{\mathcal{D}} - \underset{\sim}{\Gamma}\wedge\underset{\sim}{B}) .$$

However, it is not possible to determine $\underset{\sim}{\Gamma}\underset{\sim}{\chi}$ and $\underset{\sim}{\phi}$ separately from $\underset{\sim}{B}$.

The defect dynamics quantities, namely distortion β^i_A, velocity V^i, dislocation density α^{Ai}, dislocation current J^i_A, spin ω^i_A, bend-twist k^{Ai}, disclination density q^i, disclination current S^{Ai} are thus explicitly determined in terms of the functions χ^i, ϕ^i_a, W^α_a. Indeed, by (2.7.8), (2.7.11) and (3.7.3) we can write

$$(3.7.5a) \qquad \begin{aligned} \beta^i_A &= \partial_A\chi^i + \gamma^i_{\alpha j}\,\chi^j W^\alpha_A + \phi^i_A, \\ V^i &= \partial_4\chi^i + \gamma^i_{\alpha j}\,\chi^j W^\alpha_4 + \phi^i_4 , \end{aligned}$$

$$J^i_A = \gamma^i_{\alpha j}[W^\alpha_A \phi^j_4 - W^\alpha_4 \phi^j_A + \chi^j(\partial_A W^\alpha_4 - \partial_4 W^\alpha_A$$

(3.7.5b)
$$+ C^\alpha_{\beta\gamma} W^\beta_A W^\gamma_4)] + \partial_A \phi^i_4 - \partial_4 \phi^i_A ,$$

$$\alpha^{Ai} = \varepsilon^{ABC}\{\partial_B \phi^i_C - \partial_C \phi^i_B + \gamma^i_{\alpha j}[W^\alpha_B \phi^j_C - W^\alpha_C \phi^j_B$$

$$+ \chi^j(\partial_B W^\alpha_C - \partial_C W^\alpha_B + C^\alpha_{\beta\gamma} W^\beta_B W^\gamma_C)]\} ,$$

etc., the expressions getting more involved for the disclination quantities. However, if $W^\alpha_a = 0$, $\alpha = 1,2,3$; $a = 1,2,3,4$, i.e. $\underset{\sim}{\Gamma} = \underset{\sim}{0}$, then (3.7.1) implies that $\underset{\sim}{K} = \underset{\sim}{0}$ and $\underset{\sim}{\Omega} = \underset{\sim}{0}$ and we obtain the equations for dislocation dynamics, namely

(3.7.6)
$$\underset{\sim}{B} = d\underset{\sim}{\chi} + \underset{\sim}{\phi} ,$$
$$\underset{\sim}{D} = d\underset{\sim}{\phi} ,$$
$$\underset{\sim}{K} = \underset{\sim}{0} ,$$
$$\underset{\sim}{\Omega} = \underset{\sim}{0} .$$

It is then a simple matter to obtain from (3.7.5) the expressions for the dislocation quantities:

(3.7.7)
$$\beta^i_A = \partial_A \chi^i + \phi^i_A ,$$
$$V^i = \partial_4 \chi^i + \phi^i_4 ,$$
$$J^i_A = \partial_A \phi^i_4 - \partial_4 \phi^i_A ,$$
$$\alpha^{Ai} = \varepsilon^{ABC}(\partial_B \phi^i_C - \partial_C \phi^i_B) .$$

In this case, everything is determined by the 15 functions ϕ^i_a , χ^i , out of which 12 are independent [see(3.7.2)]. However, in what follows we will find all 27 functions, ϕ^i_a , χ^i , W^α_a , involved in the theory; namely, we will solve the complete problem with both dislocations and disclinations present. Before we proceed with the variational aspects of the Yang-Mills minimal coupling theory for the full group $G = SO(3) \triangleright T(3)$, a few sentences about the observables should be added.

The basic "observables" in defect dynamics are Burgers vectors for circuits and Frank vectors that are constructed for closed 2-dimensional surfaces. In the theory of disclination free materials, the dislocation density is defined via the Burgers vector $\{b^i\}$. However, in the full theory, namely, with disclinations present, the Frank vector $\{F^i\}$ is also important.

Let S_3 be an arbitrary 3-dimensional region of the material body whose boundary, ∂S_3 , is a closed 2-dimensional surface in E_3 . The Frank vector associated with ∂S_3 is defined by [12]

$$(3.7.8)\qquad F^i(\partial S_3) = \int_{\partial S_3} \alpha^i = \int_{S_3} d\alpha^i \ .$$

The Burgers vector associated with a circuit $\partial S_2 \subset E_3$

is given by [12]

$$b^i(\partial S_2) = \int_{\partial S_2} H_3(\alpha^i) , \tag{3.7.9}$$

where S_2 is a 2-dimensional surface and H_3 denotes the homotopy operator in E_3 (i.e. with $T = \text{const}$). In the hypersurface $T = \text{const}$ it follows from (2.7.8) that

$$\alpha^i = \mathcal{D}^i \Big|_T . \tag{3.7.10}$$

Then, by (3.7.3), (3.7.10), Stokes theorem and the fact that $\partial\partial \equiv 0$ we obtain an expression for the Frank vector in terms of the curvature $\underset{\sim}{\Theta}$ and the connection $\underset{\sim}{\Gamma}$:

$$F^i(\partial S_3) = \int_{\partial S_3} (\Theta^i_j \, \chi^j) \Big|_T + \int_{\partial S_3} (\Gamma^i_j \wedge \phi^j) \Big|_T . \tag{3.7.11}$$

Since the entries of the matrices $\underset{\sim}{\Gamma}$ and $\underset{\sim}{\phi}$ are antiexact 1-forms, (2.2.5), (3.7.3) and (3.7.10) show that we can write an explicit expression for the Burgers vector:

$$b^i(\partial S_2) = \int_{\partial S_2} \phi^i \Big|_T + \int_{\partial S_2} H_3(d\Gamma^i_j \, \chi^j \Big|_T) . \tag{3.7.12}$$

For the disclination free materials, i.e. for a pure dislocated states, $W^\alpha = 0$ implies that

$$F^i(\partial S_3) = 0 \qquad \forall\, \partial S_3 \,,$$

$$b^i(\partial S_2) = \int_{\partial S_2} \phi^i \Big|_T = \int_{S_2} d\phi^i \Big|_T \qquad \forall\, \partial S_2 \,.$$

This is a state which rises due to the breaking of the homogeneity of the translation group T(3) only and is, therefore, characterized by the "translational" dislocations described by the fields ϕ^i . However, as soon as disclinations are present, the Burgers vector is given by (3.7.12). The second term on the right-hand side of (3.7.12) then comes in due to the presence of disclinations. We have mentioned before [see Sec. 3.4] that disclinations give rise to disclination driven or "rotational" dislocations and their presence is directly confirmed in the full expression for the Burgers vector. A similar situation prevails in the expression (3.7.11) for the Frank vector. Of course, for pure dislocated material the Frank vector vanishes and need not be considered in the discussions. However, for disclinated materials, the Frank vector is nontrivial. The first term on the right-hand side of (3.7.11) reflects the effects of pure rotations and is given through the curvature acting on the state vector. The second term arises due to the interaction between translational dislocations and disclinations.

The terms $d\chi^i$ in the expressions

$$(3.7.13) \qquad B^i = d\chi^i + W^\alpha \gamma_{\alpha j}^{\ i} \chi^j + \phi^i$$

for the velocity-distortion 1-forms make contributions neither to the Burgers vector, $b^i(\partial S_2)$, for 1-circuits nor to the Frank vector, $F^i(\partial S_3)$, for 2-circuits. On the other hand, the functions $\chi^i(X^F,T)$ have been repeatedly referred to as characterizing the configuration at time T in terms of the coordinate cover (X^F) of a reference configuration. We now proceed to solidify this interpretation through a direct derivation.

Consider the line

$$(3.7.14) \qquad \eta:[0,1] \to E_4 | \bar{X}^A = \lambda X^A , \ \bar{T} = \lambda T, \ 0 \le \lambda \le 1$$

connecting the point $P_0(0,0,0,0)$ with the point $P_1(X^1,X^2,X^3,T)$. It is a trivial matter to see that the map η induces the pull back relations

$$(3.7.15) \qquad \eta^* d\bar{X}^A = X^A \, d\lambda , \ \eta^* d\bar{T} = T \, d\lambda$$

to 1-forms on $[0,1]$ for each $P_1(X^1,X^2,X^3,T)$ in E_4. Thus, if we define the quantities $x^i(X^a)$ by the line integrals

$$(3.7.16) \qquad x^i(X^a) = \int_{[0,1]} \eta^* B^i ,$$

then

$$(3.7.17) \qquad x^i(X^F,T) = \int_0^1 \{\partial_a\chi^i + W_a^\alpha\gamma_{\alpha j}^{\ i}\chi^j + \phi_a^i\}(\lambda X^F,\lambda T)X^a d\lambda \ .$$

The integration of the first term is trivial, and hence

$$(3.7.18) \qquad x^i(X^F,T) = \chi^i(X^F,T) - \chi^i(0^F,0) + \xi^i \ ,$$

where

$$(3.7.19) \qquad \xi^i = \int_0^1 X^a\{\phi_a^i + W_a^\alpha\gamma_{\alpha j}^{\ i}\chi^j\}(\lambda X^F,\lambda T)d\lambda \ .$$

However, $X^a = \lambda X^a \frac{1}{\lambda} = X^a(\lambda X^F,\lambda T)\frac{1}{\lambda}$, and hence (3.7.19) becomes

$$(3.7.20) \qquad \xi^i = \int_0^1 \{X^a\phi_a^i + X^a W_a^\alpha\gamma_{\alpha j}^{\ i}\chi^j\}(\lambda X^F,\lambda T)\frac{d\lambda}{\lambda} \ .$$

Thus, if the 1-forms (ϕ^i, W^α) satisfy the antiexact gauge conditions $X^a\phi_a^i = 0$, $X^a W_a^\alpha = 0$ at all points of E_4 , they will satisfy them at the points $(\lambda X^F,\lambda T)$ for all $\lambda\varepsilon[0,1]$ and we have $\xi^i = 0$.
Thus, (3.7.18) gives

$$(3.7.21) \qquad x^i(X^a) = \chi^i(X^a) - \chi^i(0^a) \ ;$$

that is, the mapping to the current configuration is obtained from the reference configuration by line integration of the distortion 1-forms.

For a defect free material

$$x^i(X^a) = \chi^i(X^a) - \chi^i(0^a) = \int_{P_0}^{P_1} B^i = \int_{P_0}^{P_1} d\chi^i$$

for any path connecting P_o with P_1 since $dB^i = 0$ in that case. With defects present, the only difference is that we are restricted to the lines (3.7.14) rather than arbitrary paths connecting P_o with P_1. However, the lines (3.7.14) represent absolute linear processes, if we think for the moment in terms of macroscopic thermodynamics, and hence they have a natural preference over arbitrary paths with arbitrary path parameter. Indeed, the lines (3.7.14) are the natural generalization of the "virtual quasistatic processes" that are required in order to reduce the balance of energy to the first law of thermodynamics. Viewed from this perspective, the antiexact gauge conditions, $X^a\phi^i_a = 0$, $X^a W^\alpha_a = 0$, are naturally associated with virtual quasistatic processes since they are both necessary and sufficient conditions for obtaining the evaluations (3.7.21) from the definition (3.7.16) of $x^i(X^a)$ as the line integrals of the distortion 1-forms over virtual quasistatic processes. On the other hand, if the antiexact gauge conditions are not satisfied, the definition of the linear homotopy operator H and (3.7.19) show that

(3.7.22) $\quad \xi^i = H(\phi^i + W^\alpha \gamma_{\alpha j}^{\ i} x^j)$.

It then follows directly from (3.7.18) that

$$x^i(X^a) = \chi^i(X^a) - \chi^i(0^a) + H(\phi^i + W^\alpha \gamma_{\alpha j}^{\ \ i} X^j)(X^a), \tag{3.7.23}$$

and the functions $\chi^i(X^a)$ are only "partial configuration" variables. *Single-valued mappings to current configurations thus exist for all choices of the gauge conditions.* If the antiexact gauge is used, the partial configuration variables $\chi^i(X^a)$ uniquely characterize the current configuration, but not otherwise.

3.8 Construction of the Lagrangian

The Yang-Mills minimal coupling theory assumes the existence of a Lagrangian (see 2.3). Therefore, we proceed by constructing a Lagrangian that describes the dynamics of defects. The starting point is a Lagrangian L_o that describes certain known fields and is invariant under the homogeneous action of the underlying group. For our purposes, the natural start is elasticity theory with the Lagrangian

$$L_o = T - \Psi(C_{AB}) \ ,$$

which is invariant under the homogeneous action of the group $G_o = SO(3)_o \triangleright T(3)_o$ (see 2.6). We write the function L_o for the class of materials considered here as

$$(3.8.1) \qquad L_o = \frac{1}{2} \rho_o \partial_4 \chi^i \delta_{ij} \partial_4 \chi^j - \frac{1}{8}[\lambda(e_{AB}\delta^{AB})^2 + 2\mu e_{AB}\delta^{AC} \delta^{BD} e_{CD}] \ ,$$

where e_{AB} are the components of the engineering strain

$$(3.8.2) \qquad e_{AB} = C_{AB} - \delta_{AB} = \partial_A \chi^i \delta_{ij} \partial_B \chi^j - \delta_{AB} \ ,$$

λ and μ are the Lamé constants and ρ_o is the mass density in the reference configuration (a constant for our purposes). By breaking the homogeneity of the action of the group G_o, the new compensating fields W^α_a, ϕ^i_b are introduced. According to the minimal replacement (3.5.1) and the theory given in Sections 2.3 and 3.5, the expressions (3.8.1) and (3.8.2) are replaced by

$$(3.8.3) \quad L_o = \frac{1}{2} \rho_o B^i_4 \delta_{ij} B^j_4 - \frac{1}{8}\{\lambda (E_{AB}\delta^{AB})^2 + 2\mu E_{AB}\delta^{AC} \delta^{BD} E_{CD}\}$$

and

$$(3.8.4) \quad E_{AB} = B^i_A \delta_{ij} B^j_B - \delta_{AB} ,$$

with

$$(3.8.5) \quad B^i_a = \partial_a \chi^i + \gamma^i_{\alpha j} \chi^j W^\alpha_a + \phi^i_a .$$

The Lagrangian L_o in (3.8.3) is invariant under the inhomogeneous action of the group G (see appendix 2). In addition, the minimal coupling construction requires that the original Lagrangian $L_o(B^i)$ be replaced by the new Lagrangian

$$(3.8.6) \quad L = L_o + \bar{s}\bar{L} .$$

Here, $\bar{L}$ is usually a function of the compensating fields and their derivatives, $\bar{s}$ is a coupling constant and $\bar{L}$ is required to be invariant under the inhomogeneous action of the gauge group $G = SO(3) \triangleright T(3)$. A direct analogy with Yang-Mills minimal coupling theory would lead to a construction of the Lagrangian $\bar{L}$ as a quadratic function of the components of the curvature matrix $\hat{\underset{\sim}{\Theta}}$ associated with the group $G = SO(3) \triangleright T(3)$, which is invariant under the action of the group G (see 2.3). From (3.6.8) and $\hat{\underset{\sim}{\Theta}} = d\hat{\underset{\sim}{\Gamma}} + \hat{\underset{\sim}{\Gamma}} \wedge \hat{\underset{\sim}{\Gamma}}$ we obtain an explicit expression for the curvature matrix corresponding to the connection $\hat{\underset{\sim}{\Gamma}}$ associated with a matrix representation of the group G on the affine set V of V_4;

$$\hat{\underset{\sim}{\Theta}} = \begin{pmatrix} \underset{\sim}{\Theta} & D\underset{\sim}{\phi} \\ [\underset{\sim}{0}] & 0 \end{pmatrix}$$

Here, $\underset{\sim}{\Theta}$ is the 3-by-3 curvature matrix associated with the subgroup $SO(3)$ of G. Since $\hat{\underset{\sim}{\Theta}}$ has the induced transformation law $'\hat{\underset{\sim}{\Theta}} = \underset{\sim}{M}\hat{\underset{\sim}{\Theta}}\underset{\sim}{M}^{-1}$ under the action of G, a direct calculation shows that

$$'\hat{\underset{\sim}{\Theta}} = \begin{pmatrix} '\underset{\sim}{\Theta} & 'D\underset{\sim}{\phi} \\ [\underset{\sim}{0}] & 0 \end{pmatrix} = \begin{pmatrix} \underset{\sim}{A}\underset{\sim}{\Theta}\underset{\sim}{A}^{-1} & \underset{\sim}{A}D\underset{\sim}{\phi} - \underset{\sim}{A}\underset{\sim}{\Theta}\underset{\sim}{A}^{-1}\underset{\sim}{b} \\ [\underset{\sim}{0}] & 0 \end{pmatrix};$$

that is

$$(3.8.7) \qquad '\underset{\sim}{\Theta} = \underset{\sim}{A}\underset{\sim}{\Theta}\underset{\sim}{A}^{-1}, \qquad 'D\underset{\sim}{\phi} = \underset{\sim}{A}\, D\underset{\sim}{\phi} - \underset{\sim}{A}\underset{\sim}{\Theta}\underset{\sim}{A}^{-1}\,\underset{\sim}{b}\,.$$

The underlying group of classical Yang-Mills theory is always semisimple while our underlying group is $SO(3) \triangleright T(3)$ which is not semisimple. This fact is already in evidence by the explicit occurrence of the translation $\underset{\sim}{b}$ in the induced transformation law for $D\underset{\sim}{\phi}$ given above. It follows directly from $'\hat{\underset{\sim}{\Theta}} = \underset{\sim}{M}\hat{\underset{\sim}{\Theta}}\underset{\sim}{M}^{-1}$ that $Tr('\hat{\underset{\sim}{\Theta}} \otimes '\hat{\underset{\sim}{\Theta}}) = Tr(\hat{\underset{\sim}{\Theta}} \otimes \hat{\underset{\sim}{\Theta}})$, and the representation of $\hat{\underset{\sim}{\Theta}}$ in terms of $\underset{\sim}{\Theta}$ and $D\underset{\sim}{\phi}$ gives $Tr(\hat{\underset{\sim}{\Theta}} \otimes \hat{\underset{\sim}{\Theta}}) = Tr(\underset{\sim}{\Theta} \otimes \underset{\sim}{\Theta}) = Tr('\underset{\sim}{\Theta} \otimes '\underset{\sim}{\Theta}) = F^{\alpha} \otimes F^{\beta} Tr(\underset{\sim}{\gamma}_{\alpha}\underset{\sim}{\gamma}_{\beta})$. Thus, since $Tr(\underset{\sim}{\gamma}_{\alpha}\underset{\sim}{\gamma}_{\beta})$ and $C_{\alpha\beta}$ for SO(3) agree to within a constant numerical factor, a direct analogy with Yang-Mills theory gives the gauge invariant Lagrangian

$$(3.8.8) \qquad L_2 = \frac{1}{2} C_{\alpha\beta} F^{\alpha}_{ab} g^{ac} g^{be} F^{\beta}_{ce} .$$

Here, $C_{\alpha\beta}$ are the components of the Cartan-Killing metric of the subgroup SO(3), the quantities g^{ab} are given by

$$(3.8.9) \qquad g^{AB} = -\delta^{AB}, \quad g^{44} = 1/\zeta, \quad g^{ab} = 0 \text{ for } a \neq b,$$

and $\underset{\sim}{\Theta} = F^{\alpha} \underset{\sim}{\gamma}_{\alpha}$, $F^{\alpha} = \frac{1}{2} F^{\alpha}_{ab} dX^a \wedge dX^b$, $F^{\alpha}_{ab} = \partial_a W^{\alpha}_b - \partial_b W^{\alpha}_a + C^{\alpha}_{\beta\gamma} W^{\beta}_a W^{\gamma}_b$.

The Lagrangian L_2 depends only on the W-fields and their derivatives, so there is as yet no dependence on the derivatives of the ϕ-fields. It is, however, impossible to construct a gauge invariant quantity out of the $D\phi$'s alone, as the occurence of the translation vector $\underset{\sim}{b}$ in the second of (3.8.7) clearly shows. We are thus forced back to fundamentals where the physics of defects provides the underpinning. The dislocation density and current 2-forms $\mathcal{D}^i = \alpha^i + J^i \wedge dT$ have been shown to have the evaluation $\mathcal{D}^i = D\phi^i + \Theta^i_j \chi^j = \Sigma^i$ in terms of the Cartan torsion associated with the distortion 1-forms B^i, and it is reasonable to expect the Lagrangian for a material with defects to depend on the dislocation densities and currents. On the other hand, since $D\hat{\underset{\sim}{\chi}}^T = [D\underset{\sim}{\chi}^T, 0]$, we see that $D(D\hat{\underset{\sim}{\chi}}) = \hat{\underset{\sim}{\Theta}}\hat{\underset{\sim}{\chi}} = \hat{\underset{\sim}{\Sigma}}$ and the Cartan torsion $\hat{\underset{\sim}{\Sigma}}$ has the evaluation

$$\hat{\underset{\sim}{\Sigma}}^T = [\underset{\sim}{\Sigma}^T, 0] .$$

Under the action of the gauge group G, $\hat{\underset{\sim}{\Sigma}}$ transforms according to

$$'\hat{\underset{\sim}{\Sigma}} = '\hat{\underset{\sim}{\Theta}}\ '\hat{\underset{\sim}{\chi}} = \underset{\sim}{M}\hat{\underset{\sim}{\Theta}}\underset{\sim}{M}^{-1}\underset{\sim}{M}\hat{\underset{\sim}{\chi}} = \underset{\sim}{M}\hat{\underset{\sim}{\Sigma}}$$

and consequently

$$(3.8.10) \qquad '\hat{\underset{\sim}{\Sigma}} = \begin{Bmatrix} '\underset{\sim}{\Sigma} \\ 0 \end{Bmatrix} = \underset{\sim}{M}\hat{\underset{\sim}{\Sigma}} = \begin{Bmatrix} \underset{\sim}{A}\underset{\sim}{\Sigma} \\ 0 \end{Bmatrix} = \begin{Bmatrix} \underset{\sim}{A}(D\underset{\sim}{\phi} + \underset{\sim}{\Theta}\underset{\sim}{\chi}) \\ 0 \end{Bmatrix} .$$

If the group is semisimple, $\underset{\sim}{\Sigma}$ is uniquely determined in terms of the curvature and state. On the other hand, (3.8.10) shows that $\underset{\sim}{\Sigma}$ and $\underset{\sim}{\Theta}\underset{\sim}{\chi}$ are independent whenever

$D\underset{\sim}{\phi} \neq \underset{\sim}{0}$. Thus, the independence of Cartan torsion and curvature is a direct reflection of the lack of semisimplicity of the underlying group G . Now, the fact that $\underset{\sim}{A}$ is orthogonal shows that the quadratic form $\underset{\sim}{\Sigma}^T \otimes \underset{\sim}{\Sigma}$ is positive definite and invariant under the inhomogeneous action of the gauge group G , so our goal is accomplished. Since $\underset{\sim}{\mathcal{D}} = \underset{\sim}{\Sigma}$, $\mathcal{D}^i = \frac{1}{2}\mathcal{D}^i_{ab}\, dX^a \wedge dX^b$,

(3.8.11) $$\mathcal{D}^i_{ab} = \partial_a \phi^i_b - \partial_b \phi^i_a + \gamma^i_{\alpha j}(W^\alpha_a\, \phi^j_b - W^\alpha_b\, \phi^j_a + F^\alpha_{ab}\, \chi^j) ,$$

and $\mathcal{D}^i_{ab}$ have well defined physical interpretations, the Lagrangian that describes the action of the dislocation densities and currents is

(3.8.12) $$L_1 = \frac{1}{2}\,\delta_{ij}\mathcal{D}^i_{ab}\, k^{ac}\, k^{bd}\, \mathcal{D}^j_{cd} .$$

Here, the components k^{ab} of the tensor $\underset{\sim}{k}$ are given by

(3.8.13) $$k^{AB} = -\,\delta^{AB} , \quad k^{44} = \frac{1}{y} ; \quad k^{ab} = 0 \Leftrightarrow a \neq b .$$

The Lagrangians L_1 and L_2 are not required to be Lorentz invariant, and hence the parameters ζ and y cannot be identified with the speed of light in vacuo. In fact, at this point we cannot even say whether y and ζ are positive or negative. We shall show later that it is necessary to require that both of them be positive in order to have solutions that are physically meaningful.

We note that the Lagrangian (3.8.12), with (3.8.13) can also be obtained by standard isotropy and homogeneity arguments of classical continuum theories together with the gauge invariance condition.

We can now write the Lagrangian L that describes the dynamics of defects and is invariant under the inhomogeneous action of the non semisimple gauge group $G = SO(3) \triangleright T(3)$:

$$(3.8.14) \quad L = \frac{1}{2} \rho_o B^i_4 \delta_{ij} B^j_4 + \frac{1}{8}[\lambda (E_{AB}\delta^{AB})^2 + 2\mu E_{AB}\delta^{AC} \delta^{BD} E_{CD}] - s_1 \frac{1}{2} \delta_{ij} \mathcal{D}^i_{ab} k^{ac} k^{bd} \mathcal{D}^j_{cd} - s_2 \frac{1}{2} C_{\alpha\beta} F^\alpha_{ab} g^{ac} g^{bd} F^\beta_{cd} .$$

Two facts should be noted concerning this Lagrangian. The first is that it is a natural outgrowth of the Lagrangian that describes elasticity theory with exactly four adjustable constants: two coupling constants, s_1 and s_2, and the two "propagation parameters", $\zeta = 1/g^{44}$ and $y = 1/k^{44}$. Thus, the name "minimal coupling" is clearly appropriate. The second is that L in (3.8.14) is the simplest Lagrangian that can be considered in defect dynamics. It is the sum of three Lagrangians. The first,

(3.8.15) $$ {}_{\varepsilon}L = \frac{1}{2}\rho_o B^i_4 \delta_{ij} B^j_4 - \frac{1}{8}[\lambda(E_{AB}\delta^{AB})^2 + 2\mu E_{AB}\delta^{AC}\delta^{BD}E_{CD}] , $$

is the Lagrangian that describes the action of the elastic properties of the material. The second,

(3.8.16) $$ {}_{\phi}L = -\frac{1}{2} s_1 \delta_{ij} \mathcal{D}^i_{ab} k^{ac} k^{bd} \mathcal{D}^j_{cd} , $$

is the Lagrangian that describes the action of the dislocations, and

(3.8.17) $$ {}_{W}L = -\frac{1}{2} s_2 C_{\alpha\beta} F^{\alpha}_{ab} g^{ac} g^{bd} F^{\beta}_{cd} $$

is the Lagrangian that describes the action of the disclinations. While more complicated Lagrangians ${}_{\varepsilon}L$ could be entertained in order to model intrinsic elastic nonlinearities, the Lagrangians ${}_{\phi}L$ and ${}_{W}L$ are more or less fixed by the minimal coupling construct.

The minus signs in the Lagrangians ${}_{\phi}L$ and ${}_{W}L$ are chosen for later convenience.

3.9 Notation and Useful Relations

Before we proceed with the variational aspects of the Yang-Mills minimal coupling theory for the gauge group $G = SO(3) \rhd T(3)$, we recall some useful expressions and write them in several ways.

The connection matrix $\underset{\sim}{\Gamma}$ and the curvature matrix $\underset{\sim}{\Theta}$ associated with the Lie group SO(3) are expressible in terms of the infinitesimal generating matrices $\underset{\sim}{\gamma}_\alpha$ of $SO(3)_o$ as follows (see Sec. 3.4):

$$(3.9.1) \qquad \underset{\sim}{\Gamma} = W^\alpha \underset{\sim}{\gamma}_\alpha = W^\alpha_a \, dX^a \, \underset{\sim}{\gamma}_\alpha$$

and

$$(3.9.2) \qquad \underset{\sim}{\Theta} = F^\alpha \underset{\sim}{\gamma}_\alpha = (dW^\alpha + \tfrac{1}{2} C^\alpha_{\beta\gamma} W^\beta \wedge W^\gamma) \underset{\sim}{\gamma}_\alpha ,$$

$$F^\alpha = \tfrac{1}{2} F^\alpha_{ab} \, dX^a \wedge dX^b .$$

Their components are given by

$$(3.9.3) \qquad \Gamma^i_j = W^\alpha_a \gamma^i_{\alpha j} \, dX^a ,$$

and

$$(3.9.4) \qquad \Theta^i_j = \tfrac{1}{2} F^\alpha_{ab} \gamma^i_{\alpha j} \, dX^a \wedge dX^b ,$$

$$F^\alpha_{ab} = \partial_a W^\alpha_b - \partial_b W^\alpha_a + C^\alpha_{\beta\gamma} W^\beta_a W^\gamma_b$$

respectively.

If $\underset{\sim}{\chi} \varepsilon A^{o}_{3,1}(E_4)$ is a column matrix whose entries $\{\chi^i\}$ are functions that transforms under the action of G according to $'\underset{\sim}{\chi} = \underset{\sim}{A}\underset{\sim}{\chi}$, then (2.5.27) implies

(3.9.5) $$D\underset{\sim}{\chi} = d\underset{\sim}{\chi} + \underset{\sim}{\Gamma}\underset{\sim}{\chi} ,$$

which can be written in several equivalent ways:

(3.9.6)
$$D\chi^i = d\chi^i + \Gamma^i_j \, \chi^j = d\chi^i + W^\alpha \, \gamma^i_{\alpha j} \, \chi^j = D_a\chi^i \, dX^a ,$$
$$D_a\chi^i = \partial_a\chi^i + W^\alpha_a \, \gamma^i_{\alpha j} \, \chi^j .$$

According to the minimal replacement (3.5.1), the distortion matrix of 1-forms is given by

(3.9.7) $$\underset{\sim}{B} = D\underset{\sim}{\chi} + \underset{\sim}{\phi} = d\underset{\sim}{\chi} + \underset{\sim}{\Gamma}\underset{\sim}{\chi} + \underset{\sim}{\phi} ,$$

or, in terms of its components,

(3.9.8)
$$B^i = d\chi^i + \Gamma^i_j \, \chi^j + \phi^i = B^i_a \, dX^a ,$$
$$B^i_a = \partial_a\chi^i + W^\alpha_a \, \gamma^i_{\alpha j} \, \chi^j + \phi^i_a .$$

From (3.7.1) (3.8.9) (3.9.7) and (2.3.33) it follows that the dislocation matrix can be written as

(3.9.9) $$\underset{\sim}{\mathcal{D}} = \underset{\sim}{\Theta}\underset{\sim}{\chi} + D\underset{\sim}{\phi} = D(D\underset{\sim}{\chi} + \underset{\sim}{\phi}) = D\underset{\sim}{B} = d\underset{\sim}{B} + \underset{\sim}{\Gamma} \wedge \underset{\sim}{B} .$$

The disclination density, current, the bend-twist and

spin are expressible in terms of the matrices $\underset{\sim}{\Gamma}$, $\underset{\sim}{B}$ and $\underset{\sim}{\mathcal{D}}$. From (3.7.1-2), (3.9.3) and (3.9.8) we obtain the expressions for bend-twist forms:

$$(3.9.10) \quad \underset{\sim}{K} = \underset{\sim}{\Gamma} \wedge \underset{\sim}{B} \,, \quad K^i = \Gamma^i_j \wedge B^j = \frac{1}{2}\, K^i_{ab}\, dX^a \wedge dX^b \,,$$

with

$$(3.9.11) \quad K^i_{ab} = \gamma^i_{\alpha j}\, (W^\alpha_a\, B^j_b - W^\alpha_b\, B^j_a) \,.$$

We also write relations for disclination density-current forms:

$$(3.9.12) \quad \underset{\sim}{\Omega} = d\underset{\sim}{\mathcal{D}} \,, \quad \Omega^i = d\mathcal{D}^i = \frac{1}{3!}\, \Omega^i_{abc}\, dX^a \wedge dX^b \wedge dX^c \,.$$

For convenience we introduce specific notation for the derivatives of the Lagrangian (3.8.14) with respect to the different fields. We thus define:

$$(3.9.13) \quad \bar{G}^{ab}_\alpha = \frac{\partial L}{\partial F^\alpha_{ab}} \,, \quad \bar{G}^{ab}_\alpha = - \bar{G}^{ba}_\alpha \,.$$

The quantities $\bar{G}^{ab}_\alpha$ can be regarded as components of 2-forms

$$(3.9.14) \quad \bar{G}_\alpha = \frac{1}{2}\, \bar{G}^{ab}_\alpha\, \pi_{ab} \,.$$

Furthermore, if we define the 2-forms $\bar{G}^\alpha$ by

$$(3.9.15) \quad \bar{G}^\alpha = C^{\alpha\beta}\, \bar{G}_\beta \,,$$

where $C^{\alpha\beta}$ are components of the inverse of the Cartan-Killing metric for $SO(3)$, then the matrix $\bar{\underset{\sim}{G}}$ whose entries are the 2-forms $\bar{G}^{\alpha}$ can be written in terms of the generating matrices $\underset{\sim}{\gamma}_{\alpha}$ of the group $SO(3)_{o}$ as follows

(3.9.16) $\bar{\underset{\sim}{G}} = \bar{G}^{\alpha} \underset{\sim}{\gamma}_{\alpha}$.

From (3.8.14), (3.8.19) and (3.9.13) we obtain explicit expressions for $\bar{G}^{ab}_{\alpha}$ (see(3.9.23)):

(3.9.17) $\bar{G}^{ab}_{\alpha} = - s_2 g^{ca} g^{db} C_{\alpha\beta} F^{\beta}_{cd} + R^{ab}_{i} \gamma^{i}_{\alpha j} \chi^{j}$

The derivatives of the Lagrangian L with respect to the gauge fields W^{α}_{a} are defined by

(3.9.18) $\bar{J}^{a}_{\alpha} = \left.\dfrac{\partial L}{\partial W^{\alpha}_{a}}\right|_{F^{\alpha}_{ab}}$

(i.e., for constant F^{α}_{ab}). Let $\bar{\underset{\sim}{J}}$ denote a square matrix whose entries are 3-forms $\bar{J}^{\alpha}$ defined by

(3.9.19) $\bar{J}^{\alpha} = C^{\alpha\beta} \bar{J}_{\beta}$

with

(3.9.20) $\bar{J}_{\beta} = \bar{J}^{a}_{\beta} \pi_{a} \ \varepsilon\Lambda^{3}(E_4)$.

Then, $\bar{J}^{\alpha}$ can be written in terms of the generating matrices $\underset{\sim}{\gamma}_{\alpha}$,

$$(3.9.21)\quad \underset{\sim}{\bar{J}} = \bar{J}^{\alpha}\, \underset{\sim}{\Upsilon}_{\alpha}\ .$$

Next,

$$(3.9.22)\quad R_i^{ab} = \frac{\partial L}{\partial \mathcal{D}^i_{ab}}\ ,\quad R_i^{ab} = -\, R_i^{ba}$$

and we set

$$(3.9.23)\quad R_i = \frac{1}{2}\, R_i^{ab}\, \pi_{ab}\ .$$

The 2-forms R_i can be regarded as entries of a row matrix $\underset{\sim}{R}$. Furthermore, a direct differentiation of the Lagrangian L in (3.8.14) with respect to $\mathcal{D}^i_{ab}$'s and (3.8.18) give

$$(3.9.24)\quad R_i^{ab} = -\, s_1 \delta_{ij} k^{ac}\, k^{bd} [\partial_c \phi^j_d - \partial_d \phi^j_c + \gamma^j_{\alpha k} (W^{\alpha}_c\, \phi^k_d - W^{\alpha}_d\, \phi^k_c) + \gamma^j_{\alpha k}\, F^{\alpha}_{cd}\, X^k]$$

It remains now to define the derivatives of L with respect to the distortion-velocity fields. Let

$$(3.9.25)\quad Z_i^a = \frac{\partial L}{\partial B^i_a}\ ,$$

with

$$(3.9.26)\quad Z_i^A = -\, \sigma_i^A \quad \text{and} \quad Z_i^4 = p_i\ .$$

We had before [see (2.7.13)] that

$$Z_i^a = \frac{\partial L}{\partial(\partial_a \chi^i)} ,$$

but from (3.9.8) it follows immediately that

$$\frac{\partial L}{\partial(\partial_a \chi^i)} \equiv \frac{\partial L}{\partial B_a^i} ,$$

so that the definitions (3.9.25) are consistent with the previously given definitions of stress and momentum. From (3.9.26) and (3.8.14) we obtain explicit expression for the stress,

(3.9.27) $$\sigma_i^A = \frac{1}{2}\delta_B^A\, \delta_{ij}(\partial_C \chi^j + W_C^\alpha\, \gamma_{\alpha k}^j\, \chi^k + \phi_C^j)(\lambda\delta^{BC}\, \delta^{FD}\, E_{FD} + 2\mu\delta^{RB}\, \delta^{SC}\, E_{RS})$$

and the momentum

(3.9.28) $$p_i = \rho_o \delta_{ij}(\partial_4 \chi^j + W_4^\alpha\, \gamma_{\alpha k}^j\, \chi^k + \phi_4^j) .$$

Further, we define the 3-forms Z_i by

(3.9.29) $$Z_i = Z_i^a\, \pi_a ,$$

and we denote by $\underset{\sim}{Z}$ a row matrix whose entries are the 3-forms $\{Z_i\}$.

Now, by the chain rule and by (3.9.18), (3.7.3), (3.9.22), (3.9.25) we obtain

$$(3.9.30)\quad \bar{J}^a_\alpha = \gamma^i_{\alpha j}(Z^a_i\,\chi^j + 2R^{ab}_i\,\phi^j_b)$$

which will turn out to be an important relationship between dislocations, disclinations and stress. The relation (3.9.30) can be written in the equivalent way:

$$(3.9.31)\quad \begin{cases} \bar{J}^A_\alpha = (-\sigma^A_i\,\chi^j + 2R^{Ab}_i\,\phi^j_b)\gamma^i_{\alpha j} \\ \bar{J}^4_\alpha = (p_i\chi^j + 2R^{4B}_i\,\phi^j_B)\gamma^i_{\alpha j} \;. \end{cases}$$

3.10 Variation With Respect to χ^i

The equations that determine the functions χ^i, W^α_a and ϕ^i_a arise from the requirement that the action functional

$$(3.10.1)\quad A[\chi,\phi,W] = \int_{E_4} L(B,\Theta,\mathcal{D})\pi ,$$

with the Lagrangian L given by (3.8.14), be rendered stationary in value relative to the choices of the functions χ^i, W^α_a and ϕ^i_a.

Variation of the action functional A with respect to χ^i leads to the Euler-Lagrange equations

$$(3.10.2)\quad \{E|L\}_{\chi^i} \equiv \frac{\partial L}{\partial \chi^i} - \partial_a \frac{\partial L}{\partial(\partial_a \chi^i)} = 0 .$$

From (3.8.14) together with (3.9.25) and (3.9.8) we obtain:

$$\frac{\partial L}{\partial \chi^i} = \frac{\partial L}{\partial B^j_a}\frac{\partial B^j_a}{\partial \chi^i} + \frac{\partial L}{\partial \mathcal{D}^j_{ab}}\frac{\partial \mathcal{D}^j_{ab}}{\partial \chi^i} = Z^a_j\, W^\alpha_a\, \gamma^j_{\alpha i} + R^{ab}_j\, F^\alpha_{ab}\, \gamma^j_{\alpha i}$$

and

$$\frac{\partial L}{\partial(\partial_a \chi^i)} = \frac{\partial L}{\partial B^i_a} = Z^a_i ,$$

so that the equations (3.10.2) take the form

$$(3.10.3)\quad \partial_a Z^a_i - Z^a_j\, W^\alpha_a\, \gamma^j_{\alpha i} = \gamma^j_{\alpha i}\, F^\alpha_{ab}\, R^{ab}_j .$$

By (3.9.3) and (3.9.29), equations (3.10.3) can be written in terms of the corresponding exterior forms as follows:

(3.10.4) $dZ_i + Z_j \wedge \Gamma^j_i = -\ 2R_j \wedge \Theta^j_i$.

Then, since the row matrix $\underset{\sim}{Z}$ transforms under the action of the group SO(3) according to

$$'\underset{\sim}{Z} = \underset{\sim}{Z}\underset{\sim}{A}^{-1}$$

(see appendix 3) and by (2.3.29) with $k=3$, the equations (3.10.4) assume the gauge convenient form

(3.10.5) $D\underset{\sim}{Z} = -\ 2\underset{\sim}{R} \wedge \underset{\sim}{\Theta}$.

Suppose now that $W^{\alpha}_{a} = 0$, so that there are no disclinations present; i.e. we consider a pure dislocated material. In this case, we obtain the following equations from (3.10.3),

(3.10.6) $\partial_A \sigma^A_i = \partial_4 p_i$.

These are the equations of balance of the linear momentum as are well known in classical elasticity theory. Therefore, for a pure dislocated material the equations (3.10.5) are the equations of balance of the linear momentum. However, if there are disclinations present, there are additional terms in the equations of balance, namely:

$$(3.10.7) \qquad \partial_4 p_i - \partial_A \sigma_i^A = \gamma_{\alpha i}^j (W_4^\alpha\, p_j - W_A^\alpha\, \sigma_j^A + F_{ab}^\alpha\, R_j^{ab})$$

Nevertheless, the equations (3.10.5) are the equations of balance of the linear momentum when both dislocations and disclinations are present. The terms on the right hand side of (3.10.7) are viewed as forces generated by the defects that act on the continuum as equivalent fields of body forces. The principle of "action and reaction" would suggest that the continuum should likewise give rise to forces that act on the defects. This will be shown to be the case in later sections.

Let $\delta_{\chi i}$ denote the process of variation with respect to the state variables χ^i and let $\delta\chi^j(X^A)$ denote the incremental functions that generate this variation process (i.e., $\chi^i(X^A) \mapsto \chi^i(X^A) + \varepsilon\, \delta\chi^i(X^A) + o(\varepsilon)^i)$. The χ-variation of the action (3.10.1) is then given by

$$(3.10.8) \qquad \delta_{\chi i} A = \int_{E_4} \{E|L\}_{\chi i}\, \delta\chi^i\, \pi + \int_{\partial E_4} Z_i^a\, \delta\chi^i\, \pi_a \,.$$

Satisfaction of the Euler-Lagrange equations (3.10.5) at all interior points thus gives

$$(3.10.9) \qquad \delta_{\chi i} A = \int_{\partial E_4} Z_i^a\, \delta\chi^i\, \pi_a$$

This shows that the action will be stationary with respect

to the choice of $\chi^i (\delta_{\chi^i} A = 0)$ only under the further conditions

$$(Z_i^a \, \delta\chi^i)\Big|_{\partial E_4} \pi_a\Big|_{\partial E_4} = 0 \, . \tag{3.10.10}$$

Thus, we either have to satisfy the Dirichlet data

$$\delta\chi^i\Big|_{\partial E_4} = 0 \qquad (\chi^i\Big|_{\partial E_4} \text{ specified}) \, , \tag{3.10.11}$$

the homogeneous Neumann data

$$(Z_i^a \, \pi_a)\Big|_{\partial E_4} = 0 \tag{3.10.12}$$

(traction-free spatial boundaries and zero initial and final momentum), or a combination of (3.10.11) and (3.10.12) on mutually disjoint parts of ∂E_4 . We deferr discussion of problems with inhomogeneous Neumann data (non-vanishing spatial boundary tractions) until Section 3.17.

3.11 Variation With Respect to ϕ^i_a

Consider again the action functional (3.10.1). If $\{\zeta^i\ ,\ i=1,2,3\}$ denotes an arbitrary collection of three 1-forms that induces the variation

$$(3.11.1) \quad \phi^i \rightarrow \phi^i + \varepsilon\zeta^i + o(\varepsilon)\ ,$$

then by (3.7.1) we obtain the induced variation in the $\mathcal{D}^i$'s which may be written as

$$(3.11.2) \quad \mathcal{D}^i \rightarrow \mathcal{D}^i + \varepsilon(d\zeta^i + \Gamma^i_j \wedge \zeta^j) + o(\varepsilon)\ .$$

Hence, the variation that is induced in the Lagrangian 4-form $L\pi$ is

$$(3.11.3) \quad \delta(L\pi) = \left(\frac{\partial L}{\partial \phi^i_a}\,\delta\phi^i_a + \frac{\partial L}{\partial \mathcal{D}^i_{ab}}\,\delta\mathcal{D}^i_{ab}\right)\pi\ ,$$

which, together with (3.9.22), (3.9.25) and (3.9.8), gives

$$(3.11.4) \quad \delta(L\pi) = (Z^a_i\ \delta\phi^i_a + R^{ab}_i\ \delta\mathcal{D}^i_{ab})\pi\ .$$

Written in terms of the differential forms $R_i\ ,\ Z_i\ ,\ \mathcal{D}^i$ we have

$$(3.11.5) \quad \delta(L\pi) = -\,Z_i \wedge \delta\phi^i - 2R_i \wedge \delta\mathcal{D}^i\ .$$

The variation of the form ϕ^i is $\delta\phi^i = \zeta^i$ and accordingly by (3.11.2)

(3.11.6) $\delta \mathcal{D}^i = d\zeta^i + \Gamma^i_j \wedge \zeta^j$.

Thus (3.11.5) can be written in the following way

(3.11.7) $\delta(L\pi) = - Z_i \wedge \zeta^i - 2R_i \wedge (d\zeta^i + \Gamma^i_j \wedge \zeta^j)$.

The exterior differentiation of the product of two forms gives

(3.11.8) $d(R_j \wedge \zeta^j) = dR_j \wedge \zeta^j + R_j \wedge d\zeta^j$,

so that from (3.11.7-8) we obtain

(3.11.9) $\delta(L\pi) = - (Z_j + 2R_i \wedge \Gamma^i_j - 2dR_j) \wedge \zeta^j - 2d(R_j \wedge \zeta^j)$.

By the fundamental lemma of calculus of variation for vanishing of variations on the boundary, we finally obtain the Euler-Lagrange equations with respect to the functions ϕ^i_b :

(3.11.10) $\frac{1}{2} Z_j = dR_j - R_i \wedge \Gamma^i_j$.

Written in terms of the row matrices $\underset{\sim}{Z}$ and $\underset{\sim}{R}$ and the square matrix $\underset{\sim}{\Gamma}$, equations (3.11.10) assume the equivalent form

(3.11.11) $\frac{1}{2} \underset{\sim}{Z} = d\underset{\sim}{R} - \underset{\sim}{R} \wedge \underset{\sim}{\Gamma}$.

According to appendix 3, $\underset{\sim}{R}$ transforms under the action of G by $'\underset{\sim}{R} = \underset{\sim}{R}\underset{\sim}{A}^{-1}$. Thus (2.3.29) with k=2 shows that equation (3.11.11) can also be written in the gauge convenient form

$$\frac{1}{2}\,\underset{\sim}{Z} = D\underset{\sim}{R} \,. \tag{3.11.12}$$

It is easier to see the phenomenology of the problem when the equations are written in the form (3.11.12). However, when it comes to solving specific problems, one has to write the governing equations in terms of the corresponding components, which are

$$\partial_a R^{ab}_j - \gamma^i_{\alpha j}\, W^\alpha_a\, R^{ab}_i = \frac{1}{2}\, Z^b_j \,. \tag{3.11.13}$$

These, in turn are equivalent to the system of twelve equations:

$$\begin{aligned} \partial_a R^{aB}_j - \gamma^i_{\alpha j}\, W^\alpha_a\, R^{aB}_i &= -\frac{1}{2}\,\sigma^B_j \\ \partial_A R^{A4}_j - \gamma^i_{\alpha j}\, W^\alpha_A\, R^{A4}_i &= \frac{1}{2}\, p_j \,. \end{aligned} \tag{3.11.14}$$

Satisfaction of the Euler-Lagrange equations (3.11.12) at all interior points of E_4 results in the following expression for the variation of the action integral with respect to the ϕ's (see 3.11.9)

$$\delta_{\phi i} A = \int_{\partial E_4} -2R_j \wedge \xi^j = -2 \int_{\partial E_4} R_j \wedge \delta\phi^j \,.$$

Since the ϕ's are internal variables, they can not be controlled on boundaries by external agencies. It is also clear that $\delta\phi^j\Big|_{\partial E_4} = 0$, $\phi^j\Big|_{\partial E_4}$ specified, are not gauge invariant conditions. Thus, A is rendered stationary with respect to the ϕ's only if we impose homogeneous Neumann data

$$(3.11.15) \qquad R_i\Big|_{\partial E_4} = R_i^{ab}\,\pi_b\Big|_{\partial E_4} = 0\ .$$

Since $\underset{\sim}{R}$ transforms under the action of G by $'\underset{\sim}{R} = \underset{\sim}{R}\underset{\sim}{A}^{-1}$ (see Appendix 3), the boundary conditions (3.11.15) are gauge invariant.

Let b^i be the components of the Burgers vector and t^A the components of the unit tangent vector field of a family of dislocation lines in the neighborhood of a point on a spatial boundary of a material body with unit normal field N_A . We then have $\mathcal{D}^i = \alpha^{iA}\,\mu_A = b^i\,t^A\,\mu_A$ and hence $\partial_B\phi^i_C = b^i\,t^A\,\varepsilon_{ABC}$ in a neighborhood of the spatial boundary. It then follows from (3.9.24) that

$$R_i^{AB} = -s_1\,\delta_{ij}\,\delta^{AF}\,\delta^{BD}\,b^j\,t^S\,\varepsilon_{SFD}\ ,$$

and hence

$$R_i^{AB}\,\mu_B\Big|_{\partial E_4} = -s_1\,\delta_{ij}\,\delta^{AF}\,\delta^{BD}\,b^i\,t^S\Big|_{\partial E_4}\varepsilon_{SFD}\,N_B\,dS\ ,$$

where we have set $\mu_B\Big|_{\partial E_4} = N_B\, dS$. Accordingly, the boundary conditions (3.11.15) will be satisfied only if $b^i\, t^S\Big|_{\partial E_4} N^D\, \varepsilon_{SFD} = 0$; that is, either $b^i\Big|_{\partial E_4} = 0$ or the dislocation lines are orthogonal to the boundary $(t^S\Big|_{\partial E_4} N^D\, \varepsilon_{SFD} = 0)$.

3.12 Variation With Respect to W_a^α

Let $\{\eta^\alpha, \alpha=1,2,3\}$ be a collection of three 1-forms that induces the variation

$$(3.12.1) \qquad W^\alpha \longrightarrow W^\alpha + \varepsilon\eta^\alpha + o(\varepsilon) \ .$$

Recall that [see (3.9.2)]

$$F^\alpha = dW^\alpha + \frac{1}{2} C^\alpha_{\beta\gamma} \ W^\beta \wedge W^\gamma \ ,$$

where the constants $C^\alpha_{\beta\gamma}$ are antisymmetric in the lower indices. The induced variation in the F^α's is thus given by

$$(3.12.2) \qquad F^\alpha \longrightarrow F^\alpha + \varepsilon(d\eta^\alpha + C^\alpha_{\beta\gamma} \ W^\beta \wedge \eta^\gamma) + o(\varepsilon) \ .$$

In order to obtain the Euler-Lagrange equations with respect to W^α's we proceed as in Section 3.11. Thus, the induced variation in the Lagrangian 4-form $L\pi$, with L given by (3.8.14) and use of (3.9.13-14), (3.9.18) and (3.9.20), is seen to be

$$(3.12.3) \qquad \delta(L\pi) = - 2\delta F^\alpha \wedge \bar{G}_\alpha + \delta W^\alpha \wedge \bar{J}_\alpha \ .$$

Hence, (3.12.1-2) can be used to obtain

$$(3.12.4) \qquad \delta(L\pi) = \eta^\alpha \wedge (-2d\bar{G}_\alpha + 2C^\beta_{\gamma\alpha} W^\gamma \wedge \bar{G}_\beta + \bar{J}_\alpha)$$
$$- 2d(\eta^\alpha \wedge \bar{G}_\alpha) \ .$$

Therefore, the Euler-Lagrange equations with respect to W^α's are

$$(3.12.5) \qquad d\bar{G}_\alpha - C^\beta_{\gamma\alpha} W^\gamma \wedge \bar{G}_\beta = \frac{1}{2} \bar{J}_\alpha ,$$

or, written in terms of the corresponding components:

$$(3.12.6) \qquad \partial_a \bar{G}^{ab}_\alpha - C^\beta_{\gamma\alpha} W^\gamma_a \bar{G}^{ab}_\beta = \frac{1}{2} \bar{J}^b_\alpha \ .$$

Let the quantities G^{ab}_α be defined by

$$G^{ab}_\alpha = - s_2 \frac{\partial L_2}{\partial F^\alpha_{ab}} \ .$$

Then by (3.9.13) and (3.9.22) we obtain an explicit evaluation of $\bar{G}^{ab}_\alpha$:

$$(3.12.7) \qquad \bar{G}^{ab}_\alpha = G^{ab}_\alpha + R^{ab}_i \gamma^i_{\alpha j} \chi^j \ .$$

Thus, if we substitute (3.12.7) and the expression (3.9.30) for $\bar{J}^b_\alpha$ in (3.12.6) and if we use the dislocation field equations (3.11.13) to eliminate $\partial_a R^{ab}_i$, after simple algebraic computations we obtain

$$(3.12.8) \qquad \partial_a G^{ab}_\alpha - C^\beta_{\gamma\alpha} W^\gamma_a G^{ab}_\beta = \gamma^i_{\alpha j} R^{bc}_i (\phi^j_c + \partial_c \chi^j)$$
$$+ \gamma^k_{\alpha j} \gamma^i_{\beta k} R^{cb}_i W^\beta_c \chi^j - C^\rho_{\beta\alpha} \gamma^i_{\rho j} R^{cb}_i \chi^j W^\beta_a \ .$$

The commutator of the generating matrices $\{\underset{\sim}{\gamma}_\alpha\}$ is defined by

$$[\underset{\sim}{\gamma}_\alpha, \underset{\sim}{\gamma}_\beta] = C^\rho_{\alpha\beta}\, \underset{\sim}{\gamma}_\rho ,$$

so that

$$(3.12.9) \quad C^\rho_{\beta\alpha}\, \gamma^i_{\rho j} = \gamma^i_{\beta k}\, \gamma^k_{\alpha j} - \gamma^i_{\alpha k}\, \gamma^k_{\beta j} .$$

Hence, (3.12.8-9) give the final form of the disclination equations

$$(3.12.10) \quad \partial_a G^{ab}_\alpha - C^\beta_{\gamma\alpha}\, W^\gamma_a\, G^{ab}_\beta = \frac{1}{2} J^b_\alpha .$$

Here, the 3-forms J_α are defined by

$$(3.12.11) \quad J_\alpha = J^a_\alpha\, \pi_a = 2\gamma^i_{\alpha j}\, R^{ab}_i\, B^j_b\, \pi_a = -\, 2\gamma^i_{\alpha j}\, R_i \wedge B^j .$$

The equations (3.12.10) can be written in terms of the corresponding matrices of exterior differential forms as follows

$$(3.12.12) \quad d\underset{\sim}{G} + \underset{\sim}{\Gamma} \wedge \underset{\sim}{G} - \underset{\sim}{G} \wedge \underset{\sim}{\Gamma} = \frac{1}{2} \underset{\sim}{J} .$$

Recall that $\underset{\sim}{G} = C^{\alpha\beta}\, G_\beta \underset{\sim}{\gamma}_\alpha$, $\underset{\sim}{\Gamma} = W^\alpha\, \underset{\sim}{\gamma}_\alpha$ and we define the matrix $\underset{\sim}{J}$ by

$$(3.12.13) \quad \underset{\sim}{J} = C^{\alpha\beta}\, J_\beta \underset{\sim}{\gamma}_\alpha .$$

The equation (3.12.12) assumes a gauge convenient form since the matrix $\underset{\sim}{G}$ transforms under the action of the group G by $'\underset{\sim}{G} = \underset{\sim}{A}\underset{\sim}{G}\underset{\sim}{A}$ [see app. 3 and (3.5.31) with $k = 2$] :

$$DG_{\sim} = \frac{1}{2} \underset{\sim}{J} \ . \tag{3.12.14}$$

The W's are also internal state variables so that only natural Neumann data can be applied on spatial boundaries. Since these follow from the exact part of $\delta(L\pi)$, (3.12.4) gives

$$\bar{G}^{aB}_{\alpha} \ \mu_B \Big|_{\partial E_3} = 0 \ .$$

However, (3.12.7) and (3.11.15) yield

$$\bar{G}^{aB}_{\alpha} \ \mu_B \Big|_{\partial E_3} = G^{aB}_{\alpha} \ \mu_B \Big|_{\partial E_3} + R^{aB}_{i} \ \gamma^{i}_{\alpha j} \ \chi^{j} \ \mu_B \Big|_{\partial E_3}$$

$$= G^{aB}_{\alpha} \ \mu_B \Big|_{\partial E_3}$$

and hence we have the gauge invariant boundary conditions

$$G^{aB}_{\alpha} \ \mu_B \Big|_{\partial E_3} = 0 \ . \tag{3.12.15}$$

3.13 Integrability Conditions

The equations that we have obtained by the requirement that the action functional (3.10.1) be stationary in value are:

$$\text{(3.13.1)} \quad D\underset{\sim}{Z} = -\, 2\underset{\sim}{R} \wedge \underset{\sim}{\theta} \ ,$$

$$\text{(3.13.2)} \quad D\underset{\sim}{R} = \frac{1}{2} \underset{\sim}{Z} \ ,$$

and

$$\text{(3.13.3)} \quad D\underset{\sim}{G} = \frac{1}{2} \underset{\sim}{J} \ .$$

Since the Lagrangian L in (3.8.14) is gauge invariant under the action of the group $G = SO(3) \triangleright T(3)$, equations (3.13.1-3) are gauge invariant.

The equations (3.13.1-3) cannot be solved for arbitrary matrices of 3-forms, $\underset{\sim}{Z}$ and $\underset{\sim}{J}$, and an arbitrary matrix of 1-forms, $\underset{\sim}{\Gamma}$, for we must also obtain satisfaction of the integrability condition:

$$DD\underset{\sim}{R} = \frac{1}{2} D\underset{\sim}{Z}$$

and

$$DD\underset{\sim}{G} = \frac{1}{2} D\underset{\sim}{J} \ .$$

Since $DD\underset{\sim}{R} = -\, \underset{\sim}{R} \wedge \underset{\sim}{\theta}$ and $DD\underset{\sim}{G} = \underset{\sim}{\theta} \wedge \underset{\sim}{G} - \underset{\sim}{G} \wedge \underset{\sim}{\theta}$ [see (2.3.34-35)], it follows that

(3.13.4) $$\underset{\sim}{DZ} = - 2\underset{\sim}{R} \wedge \underset{\sim}{\Theta}$$

and

(3.13.5) $$D\underset{\sim}{J} = 2(\underset{\sim}{\Theta} \wedge \underset{\sim}{G} - \underset{\sim}{G} \wedge \underset{\sim}{\Theta}) \; .$$

The integrability condition for the equation (3.13.1) is identically satisfied since any matrix of 5-forms on a 4-dimensional space vanishes identically.

For the Lagrangian L defined by (3.8.14) and by (3.12.7) we observe that

(3.13.6) $$\underset{\sim}{\Theta} \wedge \underset{\sim}{G} - \underset{\sim}{G} \wedge \underset{\sim}{\Theta} \equiv \underset{\sim}{0} \; ,$$

so that the consistency condition (3.13.5) is now simply

(3.13.7) $$D\underset{\sim}{J} = d\underset{\sim}{J} + \underset{\sim}{\Gamma} \wedge \underset{\sim}{J} + \underset{\sim}{J} \wedge \underset{\sim}{\Gamma} = \underset{\sim}{0} \; .$$

The consistency conditions (3.13.4-5) are not only a mathematical necessity, they also carry very useful information about the physics of defects.

According to (3.13.4), the balance of linear momentum equations (3.13.1) can be viewed as a direct consequence of the consistency conditions on the dislocation field equations (3.13.2). We can consider the balance of linear momentum to be a concomitant of the dislocation field equations. The next obvious question is then, that of the role of balance of moment of momentum; a question that is

By (3.9.25-26) and (3.8.14) we have

$$\gamma^i_{\alpha j}(-\ \sigma^A_i\ B^j_A + \rho_o\delta_{ik}B^k_4\ B^j_4) = 0\ ;$$

that is

$$(3.13.15)\quad \gamma^i_{\alpha j}\ \sigma^A_i\ B^j_A = 0\ .$$

Thus, the equations (3.13.11) assume the following simple form

$$(3.13.16)\quad \gamma^i_{\alpha j}\ \sigma^A_i\{\partial_A\chi^j + W^\alpha_A\ \gamma^j_{\alpha k}\ \chi^k + \phi^j_A\} = 0\ .$$

If there are no defects, so that the distortions are completely integrable, equations (3.13.16) reduce to the classical statement of balance of the moment of momentum

$$(3.13.17)\quad \sigma^A_i(\partial_B\chi^k)\ \partial_A\chi^j = \sigma^A_j(\partial_B\chi^k)\ \partial_A\chi^i,$$

where σ^A_i are components of the Piola-Kirchhoff stress tensor. With both, dislocations and disclinations present, the balance of moment of momentum equations read

$$(3.13.18)\quad \sigma^A_i(B^k_c)\ B^j_A = \sigma^A_j(B^k_c)\ B^i_A\ .$$

Notice that the moment arm is exactly the one that would be obtained by the minimal replacement (3.5.1):

$$d\underset{\sim}{\chi} \longrightarrow d\underset{\sim}{\chi} + \underset{\sim\sim}{\Gamma\chi} + \underset{\sim}{\phi} = \underset{\sim}{B}\ .$$

Equations (3.13.17) are, in fact, statements of the symmetry property of the Cauchy stress tensor in the case when there are no defects present i.e. when elasticity theory is recovered. It is inappropriate to talk about the Cauchy stress tensors when the distortions are non-integrable forms. We can, however, still identify the Cauchy stress associated with the integrable part of the distortion, $d\underset{\sim}{\chi}$. From (3.13.18) we can write

$$\text{(3.13.19)} \quad \sigma_i^A(B_c^k)\, \partial_A \chi^j - \sigma_j^A(B_c^k)\, \partial_A \chi^i = (W_A^\alpha\, \gamma_{\alpha k}^j\, \chi^k + \phi_A^j)\sigma_i^A(B_c^k) - (W_A^\alpha\, \gamma_{\alpha k}^i\, \chi^k + \phi_A^i)\sigma_j^A(B_c^k)$$

when both dislocations and disclinations are present. The terms on the left hand side of (3.13.19) are similar to those in equations (3.13.17). The terms on the right hand side of (3.13.19) can thus be viewed as couple stresses associated with defects.

Since $\sigma_i^A = -\partial L_o/\partial B_A^i$, (3.13.18) become $B_A^j\, \partial L_o/\partial B_i^A = B_A^i\, \partial L_o/\partial B_A^j$ and these equations are identically satisfied as a consequence of the fact that L_o depends on B_A^i only through the variables $E_{AB} = B_A^i\, \delta_{ij}\, B_B^j$. Accordingly, *the field equations of defect dynamics are completely integrable.*

3.14 The Field Equations of Defect Dynamics

We can now state the complete set of field equations of defect dynamics. They consist of

(1) the evolution equations

$$D\underset{\sim}{Z} = -2\underset{\sim}{R} \wedge \underset{\sim}{\Theta} , \tag{3.14.1}$$

$$D\underset{\sim}{R} = \frac{1}{2} \underset{\sim}{Z} , \tag{3.14.2}$$

$$D\underset{\sim}{G} = \frac{1}{2} \underset{\sim}{J} ; \tag{3.14.3}$$

(2) the consistency conditions

$$D\underset{\sim}{J} = \underset{\sim}{0} ; \tag{3.14.4}$$

(3) the constitutive relations

$$\underset{\sim}{J} = -2C^{\alpha\beta}\, \underset{\sim}{R}\underset{\sim}{\Upsilon}_\beta \wedge \underset{\sim}{B}\underset{\sim}{\Upsilon}_\alpha \tag{3.14.5}$$

$$Z_i = \frac{\partial L}{\partial B^i_a} \pi_a , \tag{3.14.6}$$

$$R_i = \frac{1}{2} \frac{\partial L}{\partial \mathcal{D}^i_{ab}} \pi_{ab} , \tag{3.14.7}$$

$$\underset{\sim}{G} = -\frac{1}{2} s_2 \frac{\partial L_2}{\partial F^\alpha_{ab}} \pi_{ab} C^{\alpha\beta} \underset{\sim}{\Upsilon}_\beta \tag{3.14.8}$$

$$\underset{\sim}{R} = [R_1, R_2, R_3] , \quad \underset{\sim}{B} = [B^1, B^2, B^3]^T , \tag{3.14.9}$$

$$(3.14.10) \quad L = \frac{1}{2}\,\rho_o B_4^i\, \delta_{ij} B_4^j - \frac{1}{8}[\lambda(E_{AB}\delta^{AB})^2 + 2\mu E_{AB}\delta^{AC}\,\delta^{BD}\,E_{CD}] - s_1\,\frac{1}{2}\,\delta_{ij}\mathcal{D}^i_{ab}\,k^{ac}\,k^{bd}\,\mathcal{D}^j_{cd} - s_2\,\frac{1}{2}\,C_{\alpha\beta}F^\alpha_{ab}\,g^{ac}\,g^{bd}\,F^\beta_{cd}\ ;$$

(4) <u>the defining relations</u>

$$(3.14.11) \quad \underset{\sim}{\Gamma} = W^\alpha \underset{\sim}{\gamma}_\alpha = W^\alpha_a\,\underset{\sim}{\gamma}_\alpha dX^a\ ,$$

$$(3.14.12) \quad \begin{cases} F^\alpha_{ab} = \partial_a W^\alpha_b - \partial_b W^\alpha_a + C^\alpha_{\beta\gamma}\,W^\beta_a\,W^\gamma_b\ , \\ \underset{\sim}{\Theta} = \frac{1}{2}\,F^\alpha_{ab}\,\underset{\sim}{\gamma}_\alpha dX^a \wedge dX^b\ , \end{cases}$$

$$(3.14.13) \quad \begin{cases} \mathcal{D}^i_{ab} = \partial_a\phi^i_b - \partial_b\phi^i_a + \gamma^i_{\alpha j}(W^\alpha_a\,\phi^j_b - W^\alpha_b\,\phi^j_a) \\ \qquad + F^\alpha_{ab}\,\gamma^i_{\alpha j}\,\chi^j\ , \end{cases}$$

$$(3.14.14) \quad \begin{cases} B^i_a = \partial_a\chi^i + \gamma^i_{\alpha j}\,W^\alpha_a\,\chi^j + \phi^i_a\ , \\ B^i = B^i_a\,dX^a\ ; \end{cases}$$

and

(5) <u>the defect equations</u>

$$(3.14.15) \quad \underset{\sim}{\mathcal{D}} = \underset{\sim}{\Theta}\underset{\sim}{\chi} + \underset{\sim}{\bar{\mathcal{D}}}\ , \quad \underset{\sim}{\bar{\mathcal{D}}} = D\underset{\sim}{\phi}\ ,$$

$$(3.14.16) \quad \underset{\sim}{K} = \underset{\sim}{\Gamma} \wedge \underset{\sim}{B}\ ,$$

$$(3.14.17) \quad \underset{\sim}{\Omega} = \underset{\sim}{\Theta} \wedge \underset{\sim}{B} - \underset{\sim}{\Gamma} \wedge \underset{\sim}{\mathcal{D}}\ .$$

As with any theory, the field equations are better understood if they are given in their explicit forms. When the constitutive relations and the defining relations are substituted into (3.14.1-5) and we set $\partial^A \equiv \delta^{AB}\,\partial_B$, the results are as follows:

(3.14.1a) $\partial_4 p_i - \partial_A \sigma^A_i - \gamma^j_{\alpha i}(W^\alpha_4\, p_j - W^\alpha_A\, \sigma^A_j) = \gamma^j_{\alpha i}\, F^\alpha_{ab}\, R^{ab}_j$

(3.14.2a)
$$\begin{aligned}
&\delta_{ji}\delta^{BD}(\partial^A \bar{\mathcal{D}}^i_{AD} - \tfrac{1}{y}\,\partial_4 \bar{\mathcal{D}}^i_{4D}) \\
&- \gamma^i_{\alpha j}\,\delta_{ik}\delta^{BD}(\delta^{AC}\, W^\alpha_A\, \bar{\mathcal{D}}^k_{CD} - \tfrac{1}{y}\, W^\alpha_4\, \bar{\mathcal{D}}^k_{4D}) \\
&+ \gamma^i_{\alpha k}\,\delta_{ij}\delta^{BD}[\partial^A(F^\alpha_{AD}\,\chi^k) - \tfrac{1}{y}\,\partial_4(F^\alpha_{4D}\,\chi^k)] \\
&- \gamma^i_{\alpha j}\,\gamma^k_{\beta m}\,\delta_{ik}\delta^{BD}\,\chi^m(\delta^{AC}\,F^\beta_{CD}\,W^\alpha_A - \tfrac{1}{y}\,F^\beta_{4D}\,W^\alpha_4) \\
&\qquad = \frac{1}{2s_1}\,\sigma^B_j\ ,
\end{aligned}$$

(3.14.2b)
$$\begin{aligned}
&\frac{1}{y}\,(\delta_{ji}\partial^A \bar{\mathcal{D}}^i_{A4} - \delta_{im}\delta^{AC}\,\gamma^i_{\alpha j}\,W^\alpha_A\,\bar{\mathcal{D}}^m_{C4}) \\
&+ \frac{1}{y}\,\gamma^i_{\alpha k}[\delta_{ij}\partial^A(F^\alpha_{A4}\,\chi^k) - \delta_{im}\delta^{AC}\,\gamma^m_{\beta j}\,W^\beta_A\,F^\alpha_{C4}\,\chi^k] \\
&\qquad = \frac{1}{2s_1}\,p_j\ ;
\end{aligned}$$

$$(3.14.3a)\quad \delta^{BC}[\partial^A\partial_A W^\rho_C - \partial_C\partial^A W^\rho_A - \frac{1}{\zeta}\partial_4(\partial_4 W^\rho_C - \partial_C W^\rho_4)]$$
$$+ \delta^{BC} C^\rho_{\beta\gamma}[W^\gamma_C\, \partial^A W^\beta_A + 2W^\beta_A\, \partial^A W^\gamma_C - W^\beta_A\, \delta^{EA}\, \partial_C W^\gamma_E$$
$$+ \delta^{BC} C^\rho_{\beta\gamma}\, C^\gamma_{\alpha\varepsilon}[W^\alpha_A\, \delta^{AE}\, W^\beta_E\, W^\varepsilon_C - \frac{1}{\zeta} W^\alpha_4\, W^\beta_4\, W^\varepsilon_C]$$
$$= -\frac{1}{2s_2} C^{\rho\eta} J^B_\eta ,$$

$$(3.14.3b)\quad \frac{1}{\zeta}\{\partial^A\partial_A W^\rho_4 - \partial_4\partial^A W^\rho_A + C^\rho_{\beta\gamma}(W^\gamma_4\, \partial^A W^\beta_A + 2W^\beta_A\, \partial^A W^\gamma_4$$
$$- W^\beta_A\, \delta^{AB}\, \partial_4 W^\gamma_B) + C^\rho_{\beta\gamma}\, C^\gamma_{\mu\nu}\, W^\beta_A\, \delta^{AB}\, W^\mu_B\, W^\nu_4$$
$$= \frac{1}{2s_2} C^{\rho\eta} J^4_\eta ;$$

$$(3.14.4a)\quad \gamma^i_{\alpha j}\, \sigma^A_i(\partial_A\chi^j + W^\alpha_A\, \gamma^j_{\alpha k}\, \chi^k + \phi^j_A) = 0 .$$

$$(3.14.5a)\quad J^A_\alpha = -2\gamma^i_{\alpha j}(R^{CA}_i\, B^j_C + R^{4A}_i\, B^j_4)$$
$$= -2\gamma^i_{\alpha j}\{R^{CA}_i(\partial_C\chi^j + \phi^j_C + \gamma^j_{\beta k}\, W^\beta_C\, \chi^k)$$
$$+ R^{4A}_i(\partial_4\chi^j + \phi^j_4 + \gamma^j_{\beta k}\, W^\beta_4\, \chi^k)\} ,$$

$$(3.14.5b)\quad J^4_\alpha = -2\gamma^i_{\alpha j}\, R^{C4}_i(\partial_C\chi^j + \phi^j_C + \gamma^j_{\beta k}\, W^\beta_C\, \chi^k) .$$

The constitutive relations (3.14.6-9) are explicitly given by

$$(3.14.6a)\quad p_i = \rho_o\delta_{ij}(\partial_4\chi^j + W^\alpha_4\, \gamma^j_{\alpha k}\, \chi^k + \phi^j_4) ,$$

(3.14.6b) $$\sigma^A_i = \frac{\lambda}{2} \delta^A_B \delta_{ij} (\partial_C \chi^j + W^\alpha_C \gamma^j_{\alpha k} \chi^k + \phi^j_C)(\delta^{BC} \delta^{FD} E_{FD} + \frac{2\mu}{\lambda} \delta^{RB} \delta^{SC} E_{RS}) ,$$

where the components of the engineering strain tensor are

(3.14.18) $$E_{AB} = (\partial_A \chi^i + \gamma^i_{\alpha k} W^\alpha_A \chi^k + \phi^i_A)\delta_{ij}(\partial_B \chi^j + \gamma^j_{\alpha k} W^\alpha_B \chi^k + \phi^j_B) - \delta_{AB} ,$$

and

(3.14.7a) $$R^{AB}_j = - s_1 \delta_{ji} \delta^{AC} \delta^{BD} [\partial_C \phi^i_D - \partial_D \phi^i_C + \gamma^i_{\alpha k} (W^\alpha_C \phi^k_D - W^\alpha_D \phi^k_C) + \gamma^i_{\alpha k} F^\alpha_{CD} \chi^k] ,$$

(3.14.7b) $$R^{A4}_j = \frac{1}{y} s_1 \delta_{ji} \delta^{AC} [\partial_C \phi^i_4 - \partial_4 \phi^i_C + \gamma^i_{\alpha k} (W^\alpha_C \phi^k_4 - W^\alpha_4 \phi^k_C) + \gamma^i_{\alpha k} F^\alpha_{C4} \chi^k] ;$$

(3.14.8a) $$G^{AB}_\eta = - s_2 C_{\eta\alpha} \delta^{CA} \delta^{DB} (\partial_C W^\alpha_D - \partial_D W^\alpha_C + C^\alpha_{\beta\gamma} W^\beta_C W^\gamma_D) ,$$

(3.14.8b) $$G^{A4}_\eta = \frac{1}{\zeta} s_2 C_{\eta\alpha} \delta^{CA} (\partial_C W^\alpha_4 - \partial_4 W^\alpha_C + C^\alpha_{\beta\gamma} W^\beta_C W^\gamma_4) .$$

3.15 Defects Associated with T(3) or with SO(3)

A body without disclinations is one for which there is no breaking of the homogeneity of the action of the rotation group, $SO(3)_0$. Thus, there are no compensating 1-forms $W^\alpha = W^\alpha_a(X^b)dX^a$ and no variations with respect to the W^α_a's need to be considered. With $W^\alpha = 0$, the exterior covariant derivative, D, reduces to the ordinary exterior derivative, d. The theory for a body without disclinations is thus obtained from our previous results by the replacement $D \longrightarrow d$ and ignoring all equations that were obtained in Section 3.12 by variation of the W^α's. We thus have the following field equations for materials without disclinations:

$$(3.15.1) \qquad \underset{\sim}{B} = d\underset{\sim}{\chi} + \underset{\sim}{\phi}\ , \quad Z_i = \frac{\partial L}{\partial B^i_a}\pi_a\ ,$$

$$(3.15.2) \qquad \underset{\sim}{\mathcal{D}} = d\underset{\sim}{\phi}\ , \quad R_i = \frac{\partial L}{\partial \mathcal{D}^i_{ab}}\pi_{ab}\ ,$$

$$(3.15.3) \qquad d\underset{\sim}{Z} = \underset{\sim}{0}\ ,$$

$$(3.15.4) \qquad d\underset{\sim}{R} = \frac{1}{2}\underset{\sim}{Z}\ .$$

Since (3.15.3) are the integrability conditions for (3.15.4), there are no further consistency conditions and the theory is complete. Disclinations free materials are thus significantly simpler to deal with than general materials with both disclinations and dislocations. In particular, from (3.14.6a,b), (3.14.18), (3.14.1a) and (3.14.2a,b) we have

$$(3.15.5)\quad \begin{cases} p_i = \rho_o \delta_{ij}(\partial_4 \chi^j + \phi^j_4) , \\ \sigma^A_i = \frac{1}{2} \lambda \delta^A_B \, \delta_{ij}(\partial_C \chi^j + \phi^j_C)(\delta^{BC} \, \delta^{FD} \, E_{FD} \\ \qquad + \frac{2\mu}{\lambda} \, \delta^{RB} \, \delta^{SC} \, E_{RS}) , \\ E_{AB} = (\partial_A \chi^i + \phi^i_A)\delta_{ij}(\partial_B \chi^j + \phi^j_B) - \delta_{AB} , \\ \partial_4 p_i - \partial_A \sigma^A_i = 0 ; \end{cases}$$

$$(3.15.6)\quad \begin{cases} s_1 \delta_{ji} \delta^{BD}[\partial^A(\partial_A \phi^i_D - \partial_D \phi^i_A) - \frac{1}{y} \partial_4(\partial_4 \phi^i_D - \partial_D \phi^i_4)] \\ \qquad = \frac{1}{2} \sigma^B_j , \\ \frac{s_1}{y} \delta_{ij} \partial^A(\partial_A \phi^i_4 - \partial_4 \phi^i_A) = \frac{1}{2} p_j . \end{cases}$$

If displacements $u^i(X^b)$ are introduced by the substitution

$$(3.15.7)\quad \chi^i(X^b) = \delta^i_A \, X^A + u^i(X^b) ,$$

then from (3.15.5) it follows that

$$(3.15.8)\quad E_{AB} = (\delta^i_A + \partial_A u^i + \phi^i_A)\delta_{ij}(\delta^j_B + \partial_B u^j + \phi^j_B) - \delta_{AB}$$

$$(3.15.9) \qquad p_i = \rho_o \delta_{ij} (\partial_4 u^j + \phi_4^j) \ .$$

Hence, we have 15 variables $(u^i , \phi_A^i , \phi_4^i)$ and 15 equations, (3.16.6) and $(3.15.5)_4$, for their determination. However, not all of the ϕ_a^i's are independent, they have to satisfy the 3 antiexact conditions:

$$(3.15.10) \qquad X^A \phi_A^i + T\phi_4^i = 0 \ ,$$

so that it would appear that the system of equations is overdetermined. Fortunately, the equations $dR_i = \frac{1}{2} Z_i$ (i.e. (3.15.6)) imply the 3 equations $dZ_i = 0$ (i.e. $(3.15.5)_4$), so that we end up with the same number of equations as the number of independent unknowns.

Although the antiexact gauge condition (3.15.10) is fundamental to the theory presented here, the fact that the evolution equations are gauge invariant allows us to apply any gauge that is convenient in their analysis. If we impose the "pseudo Lorentz" gauge condition

$$(3.15.11) \qquad 0 = \partial^A \phi_A^i - \frac{1}{y} \partial_4 \phi_4^i = \partial_a (k^{ab} \phi_b^i) \ ,$$

the system (3.15.6) reduces to

$$(3.15.12) \qquad \begin{cases} s_1 \delta_{ji} \delta^{BD} (\partial^A \partial_A \phi_D^i - \frac{1}{y} \partial_4 \partial_4 \phi_D^i) = \frac{1}{2} \sigma_j^B \ , \\ \frac{s_1}{y} \delta_{ji} (\partial^A \partial_A \phi_4^i - \frac{1}{y} \partial_4 \partial_4 \phi_4^i) = \frac{1}{2} p_j \ . \end{cases}$$

By inspection we see that the system (3.15.12) is uniformly hyperbolic provided

$$(3.15.13) \qquad y > 0 .$$

Thus, since the balance of linear momentum equations, $(3.15.5)_4$, are also uniformly hyperbolic, the Cauchy problem for materials without disclinations is well posed.

Suppose now that only the homogeneity of the action of the rotation group $SO(3)_o$ is broken. This is the original Yang-Mills construct since $SO(3)$ is semi-simple. Accordingly, we put $s_1 = 0$, $\underset{\sim}{\phi} = 0$, and (3.8.14) becomes

$$(3.15.14) \qquad L = L_o - s_2 L_2 .$$

By variation of the action functional (3.10.1), for the Lagrangian given by (3.15.14), with respect to the external fields χ^i we obtain the balance of linear momentum equations

$$(3.15.15) \qquad D\underset{\sim}{Z} = \underset{\sim}{0} .$$

Variation with respect to the compensating fields W_a^α yields the field equations

$$(3.15.16) \qquad D\underset{\sim}{\bar{G}} = \frac{1}{2} \underset{\sim}{\bar{J}} .$$

From (3.12.7) and (3.9.30), since $\underset{\sim}{\phi} = \underset{\sim}{0}$ and $\underset{\sim}{R} = \underset{\sim}{0}$, the constitutive relations are given by

$$2\bar{\underset{\sim}{G}} = - s_2 C^{\alpha\beta} \frac{\partial L_2}{\partial F^{\alpha}_{ab}} \pi_{ab} \underset{\sim}{Y}_{\beta} ,$$

$$\bar{\underset{\sim}{J}} = C^{\alpha\beta} \underset{\approx}{Z}\, \underset{\sim}{Y}_{\alpha} \underset{\approx}{X}\, \underset{\sim}{Y}_{\beta} . \tag{3.15.17}$$

3.16 Momentum-Energy Tensors and Forces

Because the theory derives from a variational principle, it has a well defined momentum-energy tensor, T^a_b, that can be expressed directly in terms of the Lagrangian, derivatives of the Lagrangian, and the field variables. The total Lagrangian, $L = L(\chi, \phi, W, \partial_a\chi, \partial_a\phi, \partial_a W)$ for materials with defects is given by (3.8.14) and admits the additive decomposition

$$L = L_o - s_1 L_1 - s_2 L_2 . \tag{3.16.1}$$

Since the components of the momentum-energy tensor are given by

$$T^a_b = \frac{\partial L}{\partial(\partial_a\chi^i)}\partial_b\chi^i + \frac{\partial L}{\partial(\partial_a\phi^i_e)}\partial_b\phi^i_e + \frac{\partial L}{\partial(\partial_a W^\alpha_e)}\partial_b W^\alpha_e - \delta^a_b L , \tag{3.16.2}$$

the additive decomposition of L leads directly to the additive decomposition

$$T^a_b = T^a_{ob} - T^a_{1b} - T^a_{2b} . \tag{3.16.3}$$

Here,

$$T^a_{ob} = \partial_b\chi^i \frac{\partial L_o}{\partial(\partial_a\chi^i)} - \delta^a_b L_o \tag{3.16.4}$$

is the momentum-energy tensor of the "elastic response",

$$(3.16.5) \qquad T^a_{1b} = s_1 \partial_b \phi^i_e \frac{\partial L_1}{\partial(\partial_a \phi^i_e)} + s_1 \partial_b W^\alpha_e \frac{\partial L_1}{\partial(\partial_a W^\alpha_e)} - s_1 \delta^a_b L_1$$

is the momentum-energy tensor of the "dislocation response", and

$$(3.16.6) \qquad T^a_{2b} = s_2 \partial_b W^\alpha_e \frac{\partial L_2}{\partial(\partial_a W^\alpha_e)} - s_2 \delta^a_b L_2$$

is the momentum-energy tensor of the "disclination response". By (3.9.13), (3.9.22) and (3.9.25), these expressions can be rewritten as follows:

$$T^a_{ob} = Z^a_i \, \partial_b \chi^i - \delta^a_b L_o ,$$

$$(3.16.7) \qquad T^a_{1b} = - 2R^{ae}_i (\partial_b \phi^i_e + \gamma^i_{\alpha j} \chi^j \partial_b W^\alpha_e) - s_1 \delta^a_b L_1 ,$$

$$T^a_{2b} = - 2G^{ae}_\alpha \, \partial_b W^\alpha_e - s_2 \delta^a_b L_2 .$$

Since any solution of the field equations (3.14.1-3) gives

$$\partial_a T^a_b = 0 ,$$

the expressions (3.16.7) lead to the "balance of force and energy"

$$(3.16.8) \qquad F_{ob} = F_{1b} + F_{2b} ,$$

where

$$(3.16.9) \qquad F_{ob} = \partial_a T^a_{ob} \ , \quad F_{1b} = \partial_a T^a_{1b} \ , \quad F_{2b} = \partial_a T^a_{2b} \ .$$

Simple calculations and the field equations yield explicit evaluation of the forces and energies:

$$(3.16.10) \qquad F_{oa} = \sigma^B_i \ \partial_a \phi^i_B - p_i \partial_a \phi^i_4 + \gamma^i_{\alpha j} \ \chi^j (\sigma^B_i \ \partial_a W^\alpha_B - p_i \partial_a W^\alpha_4) + \gamma^i_{\alpha j} \ F^\alpha_{cb} \ R^{cb}_i \ \partial_a \chi^j \ ,$$

$$(3.16.11) \qquad F_{1a} = \sigma^B_i \ \partial_a \phi^i_B - p_i \partial_a \phi^i_4 - \gamma^i_{\alpha j} \ \chi^j \ Z^b_i \ \partial_a W^\alpha_b - \gamma^k_{\alpha i} \ R^{ec}_k \{ 2W^\alpha_e (\partial_a \phi^i_c + \gamma^i_{\beta j} \ \chi^j \ \partial_a W^\beta_c) - \partial_a (W^\alpha_e \ \phi^i_c - W^\alpha_c \ \phi^i_e) + \partial_e (2\chi^i \ \partial_a W^\alpha_c) - \partial_a (\chi^i \ F^\alpha_{ec}) \} \ ,$$

$$(3.16.12) \qquad F_{2a} = - J^b_\alpha \ \partial_a W^\alpha_b + C^\eta_{\beta\gamma} [2C_{\alpha\eta} C^{\gamma\delta} \ G^{bc}_\delta \ W^\beta_b \ \partial_a W^\alpha_c + G^{bc}_\eta \ \partial_a (W^\beta_b \ W^\gamma_c)] \ .$$

In the absence of disclinations, (i.e. when $W^\alpha_a = 0$) and for a static problem (i.e. when $\phi^i_4 = 0$, $p_i = 0$) the dislocation force $\{F_{1A}\}$ reproduces the Peach-Koehler force [11] for a disclination free distribution of dislocations. Let $\{b^i\}$ be a Burgers vector defined by (3.7.12) and $\{t^A\}$ a unit tangent vector to a dislocation line. Then, for the static case

$$(3.16.13) \qquad \mathcal{D}^i|_T = \alpha^{iA} \ \mu_A = b^i t^A \ \varepsilon_{ABC} dX^B \wedge dX^C \ .$$

By $(3.7.5)_4$ we immediately obtain

$$(3.16.14) \qquad \partial_B \phi^i_C = b^i t^A \varepsilon_{ABC} .$$

Hence, (3.16.11) and (3.16.14) give an expression for Peach-Koehler force

$$(3.16.15) \qquad F_{1B} = \sigma^C_i \partial_B \phi^i_C = - \varepsilon_{BCD} t^C (b^i \sigma^D_i) .$$

In the presence of disclinations, equations (3.16.10-12) give a full description of the systems of forces that act on the elastic body, on the dislocations, and on the disclinations. Particular note should be made of the fact that the balance of force and energy equations, (3.16.8), establish exact dynamic action and reaction relationships. For example, in the static, disclination-free case just considered, (3.16.10) and (3.16.11) show that the elastic medium exerts a force on the dislocation that is equal and opposite to the force exerted by the dislocation on the elastic medium.

When the index a has the value 4, (3.16.10) gives an explicit evaluation of the elastic excess energy. A significant simplification of this expression obtains when we restrict attention to materials without disclinations. In this case the W^α_a fields are ignored from the beginning and the governing field equations are those given at the beginning of Section 3.15. The elastic excess energy then assumes the form

$$(3.16.16)\quad F_{o4} = \sigma_i^B\, \partial_4\phi_B^i - p_i\, \partial_4\phi_4^i \,.$$

If we further assume that the antiexact gauge is used, the distortion and distortion velocity are given by

$$(3.16.17)\quad B_A^i = \partial_A\chi^i + \phi_A^i \,,\quad V^i = B_4^i = \partial_4\chi^i + \phi_4^i \,.$$

and $\chi^i = \delta_A^i\, X^A + u_{total}^i$. However, the standard decomposition into elastic and plastic parts that underlies plasticity theory gives $\partial_A\chi^i = B_A^i + \overset{P}{B}{}_A^i$, $\partial_4\chi^i = V^i + \overset{P}{V}{}^i$. A comparison with (3.16.17) thus shows that the plastic distortion and the plastic velocity have the evaluations

$$(3.16.18)\quad \overset{P}{B}{}_A^i = -\,\phi_A^i \,,\quad \overset{P}{V}{}^i = -\,\phi_4^i \,,$$

respectively. We now substitute (3.16.18) into (3.16.16) to obtain

$$(3.16.19)\quad F_{o4} = -\,\sigma_i^A\, \partial_4\overset{P}{B}{}_A^i + p_i\, \partial_4\overset{P}{V}{}^i$$

Now, $\sigma_i^A\, \partial_4\overset{P}{B}{}_A^i$ is simply the rate at which work is done by the plastic distortion. In quasistatic processes where $p_i\, \partial_4\overset{P}{V}{}^i$ can be neglected, Drucker's postulate ([36], Section 18) that the rate of work done by the plastic distortion is nonnegative gives the result that the elastic excess energy is nonpositive. Thus, for this situation, the elastic response always looses energy to the dislocation fields.

Although this argument in no way proves or disproves Drucker's postulate, it does provide an immediate and useful interpretation of the elastic excess energy F_{o4} . Further, consideration is not restricted to the quasistatic case; indeed, (3.16.19) provides an exact measure of the energy transfer mechanism between the elastic response and the dislocation field that may be interpreted as the dynamic generalization of the fundamental expression $\sigma_i^A \, \partial_4 \overset{P}{B}{}_A^i$ characteristic of quasistatic processes. It is clearly inappropriate to require F_{o4} to be nonpositive at every space-time point in E_4 , as is evidenced by problems in which there are displacement and/or distortion waves. An obvious candidate in the general case would be satisfaction of the condition

$$\int_{-\infty}^{T} \left(\int_{B} F_{o4} \mu \right)^{\wedge} dT \leq 0 \ ,$$

since it would imply the existence of the essential aspects of plasticity; namely, irreversibility. In any event, it should be clear that the theory provides a positive expectation of describing the essential aspects of phenomenological plasticity in terms of the dynamics of defects in materials.

3.17 Discussion

The development of the theory given in previous sections started with the observation that the Cauchy problem for the conventional equations of defect dynamics is ill-posed. Combinations of the Yang-Mills minimal coupling theory, the conventional equations for defect dynamics and the Cartan structure equations enabled us to obtain a complete field theory for materials with dislocations and disclinations. As noted, the Yang-Mills theory consists of two parts: minimal replacement and minimal coupling. A direct consequence of the minimal replacement construct is that the deformation gradients $\partial_A \chi^i$ and Newtonian velocity $\partial_4 \chi^i$ are replaced by the distortions B^i_A and distortion velocity V^i as given in (3.7.5). This replacement is a deduced consequence of gauge invariance rather than an imposed condition. In recent works, the replacement of deformation gradients by distortions is usually argued from the point of view that defect dynamics should be able to describe plasticity theory. Accordingly, the integrable displacements are simply replaced by nonintegrable distortions in order to prevent a "stress response" from "plastic strain". This is not a justifiable argument, since the theory of plasticity is not, as yet, derived from the theory of defect dynamics. Also, with the Newtonian velocity $\partial_4 \chi^i$ replaced by the distortion velocity V^i, the form of Newton's laws in defect dynamics is equivocated.

We have been able to show solely from the minimal replacement construct, that a pure disclinated material cannot exist [see Sec. 3.4]. The presence of disclinations implies the presence of dislocations. This allows us to distinguish the so called disclination driven dislocations (or rotational dislocations) from translational dislocations. The latter arise due to the inhomogeneous action of the translation group $T(3)$ and are described by the fields ϕ^i_a. The inhomogeneous action of the rotation group $SO(3)$ gives rise not only to disclinations but also to rotational dislocations. One can thus view a total rotation as a composition of a genuine or eigen rotation and a rotation that can be realized by successive infinitesimal translations. This situation has a direct analog with spin and orbital rotation in quantum mechanics. Therefore by breaking the homogeneity of the action of the rotation group $SO(3)$ two kinds of defects arise; pure rotational ones (disclinations), and the associated translational defects (rotational dislocations). This effect is also directly evident in the expressions for Burgers and Frank vectors (3.7.11-12). The first term

$\int_{\partial S_2} \phi^i|_T$, is the Burgers vector due to the pure translational dislocations, while the second term, $\int_{\partial S_2} H_3(d\Gamma^i_j\, \chi^j|_T)$, arises solely due to the rotational dislocations. Similarly, the Frank vector (3.7.11) consists of two parts; one that is due to the presence of disclinations, $\int_{\partial S_3} (\Theta^i_j\, \chi^j)|_T$, and the term $\int_{\partial S_3} (\Gamma^i_j \wedge \phi^j)|_T$ that arises due to the dislocations. Perhaps one could view the rotational dislocations as the ones related to dislocation loops and the disclinations as characteristic of the presence of Frank-Read sources. This view is strictly conjectural at this point; a proof, if forthcoming, is left for future work.

The second part of Yang-Mills theory, the minimal coupling construct, enables us to construct a Lagrangian that describes materials with dislocations and disclinations.

Defect dynamics is an outgrowth of elasticity theory. Therefore, it is only natural to start with a Lagrangian that describes the elastic behaviour of materials. The Lagrangian L_o that we have chosen as a point of departure is the simplest one. The theory could also be built up on the premise of a more complex Lagrangian L_o . In such instance we would expect to obtain theories that describe different and certainly more complex phenomena. At the same time, the Lagrangians L_1 and L_2 , given by (3.8.12) and (3.8.8), respectively, are essentially fixed by the minimal coupling construct. One could, of course, consider additional terms in attempts to model explicit interactions between dislocations and disclinations, but this would also lead to significantly greater complexities and vagaries.

With the Lagrangian (3.8.14), the Euler-Lagrange equations that we obtain by variational methods are given by (3.14.1-3):

$$D\underset{\sim}{Z} = - 2\underset{\sim}{R} \wedge \underset{\sim}{\Theta} \; ; \quad D\underset{\sim}{R} = \frac{1}{2} \underset{\sim}{Z} \; ; \quad D\underset{\sim}{G} = \frac{1}{2} \underset{\sim}{J} \; .$$

The first set (3.14.1) are balance of linear momentum equations. In the absence of disclinations they reduce to the classical statements of balance of the linear momentum

$$\partial_A \sigma_i^A = \partial_4 p_i \quad (\text{i.e.} \quad d\underset{\sim}{Z} = \underset{\sim}{0}) \; .$$

However, the presence of defects gives rise to forces so that the balance of the linear momentum equations are no longer homogeneous in the derivatives. At the same time Newton's third law would suggest the presence of forces that act on disclinations. This is exactly what we obtained in Section 3.16. From (3.16.8) and (3.16.10-12) we can write a specific statement of balance of forces,

$$(3.17.1)\qquad F_{oA} = F_{1A} + F_{2A} ,$$

$$(3.17.2)\qquad F_{oA} = [\sigma^B_i \, \partial_A \phi^i_B - p\, \partial_A \phi^i_4] + [\gamma^i_{\alpha j}\, \chi^j (\sigma^B_i \, \partial_A W^\alpha_B - p_i \partial_A W^\alpha_4) + \gamma^i_{\alpha j}\, F^\alpha_{cb}\, R^{cb}_i \, \partial_A \chi^j] ,$$

$$(3.17.3)\qquad F_{1A} = - Z^b_i \, \partial_A \phi^i_b - [Z^b_i \, \gamma^i_{\alpha j}\, \chi^j \partial_A W^\alpha_b + \gamma^k_{\alpha i}\, R^{ec}_k (2W^\alpha_e (\partial_A \phi^i_c + \gamma^i_{\beta j}\, \chi^j \, \partial_A W^\beta_c) + \partial_A (W^\alpha_e \, \phi^i_c - W^\alpha_c \, \phi^i_e) - \partial_e (2\chi^i \, \partial_A W^\alpha_c) + \partial_A (\chi^i \, F^\alpha_{ec}))]$$

$$(3.17.4)\qquad F_{2A} = - J^b_\alpha \, \partial_A W^\alpha_b + C^\eta_{\beta\gamma} [2C_{\alpha\eta} C^{\gamma\delta}\, G^{bc}_\delta \, W^\beta_b \, \partial_A W^\alpha_c + G^{bc}_\eta \, \partial \, (W^\beta_b \, W^\gamma_c)] .$$

It is to be noted that the elastic excess forces $\{F_{o\Lambda}\}$ exactly balance the forces exerted by the medium on the dislocations, (3.17.3), and on the disclinations, (3.17.4). In the absence of disclinations (i.e. when $W^\alpha_a = 0$) , it is evident that the forces acting on dislocations, $\{F_{1A}\}$ given by (3.17.3), are precisely equilibrated by the elastic excess forces $\{F_{oA}\}$ [see (3.17.2)]. We also showed that in this case, the forces (3.17.3) are related to the

Peach-Koehler forces acting on uniform fields of dislocations.

With both dislocations and disclinations present, the situation is more complex. The defect forces (3.17.3-4) can be considered as reaction forces to those that occur on the right-hand side of the balance of the linear momentum equations. We have already identified the first terms in (3.17.3) as forces exerted by the medium on the dislocations. However, in the presence of disclinations we distinguish between translational and rotational dislocations. Therefore, those first terms are, in fact, forces on translational dislocations, while the terms in the brackets in (3.17.3) can be viewed as forces exerted by the medium on the rotational dislocations.

Similarly the elastic excess forces in (3.17.2) have two contributions, one due to the dislocations and the other due to the presence of disclinations. We also note that the elastic excess forces vanish in the absence of defects $(\phi^i_a = 0 \;, W^\alpha_a = 0)$, as is to be expected.

If we take $a = 4$ in (3.16.8) and (3.16.10-12) we obtain the balance of energy exchange equation,

$$(F_o)_4 = (F_1)_4 + (F_2)_4 \;.$$

Here $(F_o)_4$ is elastic excess stored energy in defected

materials and $(F_1)_4$, $(F_2)_4$ are exchange energies due to the dislocations and disclinations, respectively.

Now, we return to the evolution equations of defect dynamics.

Equations (3.14.2) are viewed as balance of dislocation equations. From these we see that whenever there are nonzero stresses, the dislocation field variables are nontrivial. We may thus say that stresses drive dislocations. In fact, we shall see in the next chapter that dislocations in the r-order approximation are always driven by stresses determined in the (r-1)-order approximation. The integrability conditions for the balance of dislocation equations reproduce the equations (3.14.1) of the balance of linear momentum which are explicitly included as field equations of defect dynamics.

From the balance of disclination equations (3.14.3) we conclude that dislocations and distortions both serve as sources for disclinations; we say that disclinations are driven by the fields $J^a_\alpha = J^a_\alpha(R,B)$. Necessary conditions for zero disclinations are thus

(3.17.5) $\quad \underset{\sim}{J} = \underset{\sim}{0}$.

In this case, the disclination fields W^α are trivial since they satisfy homogeneous differential equations with homogeneous initial data. By (3.12.13), (3.17.5) becomes a system of equations that the ϕ-fields have to satisfy

$$R_i^{ab} \, \gamma_{\alpha j}^i \, B_b^j = 0 \tag{3.17.6}$$

Hence, only those solutions of (3.14.2) that satisfy conditions (3.17.6) are valid in the absence of disclinations.

The integrability conditions (3.14.4) for the balance of disclination equations give the equations of balance of moment of momentum. This is a very important aspect of the theory. The balancing of the moment of momentum has usually been neglected in defect dynamics. The argument was that as a physical law that has to be satisfied in any situation is satisfied here also. However, in our theory, balance of moment of momentum equations are integrability conditions for the equations that describe the internal rotational degrees of freedom and are thus explicitly obtained. In the absence of defects, our results reduce exactly to the classical statement of balance of the moment of momentum (3.13.18)

$$\sigma_i^A \, \partial_A \chi^j = \sigma_j^A \, \partial_A \chi^i \, ,$$

where σ_i^A are the components of the Piola-Kirchhoff stress tensor. When dislocations and disclinations are present, (3.13.18) are replaced by (3.13.19)

$$\sigma_i^A(B_c^k)B_A^j = \sigma_j^A(B_c^k)B_A^i \ ;$$

the distortion B_A^j is the correct moment arm. The terms

$$\sigma_i^A(B_c^k)W_A^\alpha \ \chi^k \ \gamma_{\alpha k}^j$$

and

$$\sigma_i^A(B_c^k)\phi_A^j$$

that occur in (3.13.19) (recall that $B_A^j = \partial_A \chi_A^i + \phi_A^j + W_A^\alpha \ \gamma_{\alpha k}^j \ \chi^k$) can thus be viewed as couple stresses due to disclinations and due to dislocations, respectively.

We have seen how useful the Yang-Mills theory is in the development of the complete field theory of defect dynamics. Since there is a strong interaction between these two theories, there is a hope that a better understanding of defects in materials can bring more light into particle physics where Yang-Mills theory originated. For very large disclination energy density coefficient, s_2 , the balance of disclinations equations $D\underset{\sim}{G} = \underset{\sim}{0}$ coincide with the free Yang-Mills equations in particle physics. We remind the reader that the underlying group for the original Yang-Mills theory is the unitary group SU(2) . For purposes of argument we can consider a subgroup of the full underlying group of the defect dynamics, $G = SO(3) \triangleright T(3)$, namely, the semi-simple rotation group SO(3). Since the underlying groups SO(3) and SU(2) have isomorphic Lie algebras, the equations of Yang-Mills free fields and of disclination dynamics coincide, and the known solutions of one theory can be used in the other. The known static solution of the Yang-Mills equation is the Yang-Wu solution [16] and indeed, when used in defect dynamics, it reveals interesting results. In the same manner, possible new solutions in defect dynamics may clarify certain aspects of particle physics.

The theory given in this work is the theory of the continuous distribution of defects in materials. Therefore, it is not necessary that it will give answers to the problems with single dislocation and single disclination. The transition from the continuum theory of defects to the discrete theory is very complex. At this point it is not clear what kind of singularities the internal field variables ϕ^i_a and W^α_a should have in order to be able to describe single defects.

The field variables ϕ^i_a and W^α_a are internal field variables. They are therefore not controllable to the same extent as the external field variables χ^i. Further, since the reference configuration is taken to be defect free, the Cauchy data for ϕ^i_a and W^α_a all vanish. Accordingly, static problems must be viewed as "large time" limits of Cauchy problems. Independence of the history of the controllable external data may not be assumed *a priori*.

The fact that the field variables ϕ^i_a and W^α_a are internal variables is of fundamental significance from an entirely different point of view. It seems to be a widely accepted that field equations derived from a variational principal can not model processes that are dissipative in the macroscopic thermodynamic sense. In point of fact, it has been shown [41] that there are as many independent modes of dissipation as there are independent internal variables. For the theory presented here, what happens is that the elastic mechanical energy is partially redistributed into the internal defect degrees of freedom ϕ^i_a and W^α_a where it becomes irricoverable. Thus, although we have derived the theory from a variational principle beginning with elasticity theory, the resulting field equations describe intrinsically dissipative mechanisms as soon as the fields ϕ^i_a and or W^α_a become nontrivial.

3.18 Disclination Free Finite Bodies with Applied Surface Tractions

The theory developed up to this point is basically that of material bodies of infinite spatial extent, as is evident from the form of the action integral

$$A[\chi, \phi, W] = \int_{B\times[T_o,T_1]} (L_o - s_1 L_1 - s_2 L_2)\pi \ ;$$

that is, there are no boundary integrals that account for the work done by the external environment on the boundary of the body. Although we can use the above action integral to model finite bodies with either Dirichlet or homogeneous Neumann data, interaction of the body with the external environment are not accomodated and hence the theory is incomplete.

Most problems of practical interest do not involve disclinations, so we will confine the discussion to disclination free bodies. In the absence of interactions with the external environment, disclination free bodies are modeled by the action integral

$$(3.18.1)\quad A[\chi, \phi] = \int_{E_4} (L_o - s_1 L_1)\pi \ .$$

For a finite body B, this becomes

$$(3.18.2)\quad A[\chi, \phi] = \int_{T_o}^{T_1} \int_B (L_o - s_1 L_1)\pi \; .$$

The accomodation of interactions with the external enviro-ment is achieved by adding the integral of an exact 4-form since this can be converted to an integral over the boundary by Stokes' theorem. Clearly, an arbitrary exact 4-form will not due since it will result in changes in the Euler-Lagrange equations at interior points. The exact 4-form must leave the Euler-Lagrange equations of classic elasticity invariant since the whole theory started with the action integral

$$(3.18.3)\quad A_{el}[\chi] = \int_{T_o}^{T_1} \int_B L_o \pi$$

and then obtained (3.18.2) by the Yang-Mills minimal replacement and minimal coupling arguments. Exact 4-forms with this property are known as elements of the *null class* of the Euler-Lagrange operator [3, 39] and are always present in problems with inhomogeneous Neumann data [40].

Starting with the action integral (3.18.3) of classical elasticity, we make the replacement

$$(3.18.4)\quad L_o \pi \mapsto L_o \pi + d(\chi^i \, dQ_i) = L_o \pi + d\chi^i \wedge dQ_i$$

where Q_i , $i = 1,2,3$, are three 2-forms on E_4 that do not depend on the state variables χ^i . Thus, we may write

$$Q_i = \frac{1}{2} Q_i^{ab}(\chi^e)\pi_{ab} , \quad Q_i^{ab} + Q_i^{ba} = 0$$

since $\{\pi_{ab}\}$ is a basis for $\Lambda^2(E_4)$. It is then a simple matter to see that

$$dQ_i = \partial_b Q_i^{ba} \pi_a$$

and hence

$$d\chi^i \wedge dQ_i = \partial_a \chi^i \partial_b Q_i^{ba} \pi .$$

The replacement (3.18.4) thus leads to the following replacement for the Lagrangian function of classic elasticity:

$$(3.18.5) \quad L_o \mapsto L_o + \partial_a \chi^i \partial_b Q_i^{ba} .$$

Thus, when the minimal replacement and minimal coupling constructs are used, (3.18.2) is replaced by the new action integral

$$(3.18.6) \quad A[\chi, \phi] = \int_{T_o}^{T_1} \int_B (L_o + B_a^i \partial_b Q^{ba} - s_1 L_1)\pi .$$

An examination of the field equations (3.15.1)-(3.15.4) for bodies without disclinations shows that the only change that results from replacing (3.18.2) by (3.18.6) is

$$(3.18.7)\quad Z_i \mapsto \hat{Z}_i = Z_i + \partial_b Q_i^{ba}\,\pi_a = Z_i + dQ_i$$

We thus have the field equations

$$(3.18.8)\quad \underset{\sim}{B} = d\underset{\sim}{\chi} + \underset{\sim}{\phi}\ ,\qquad \hat{Z}_i = \frac{\partial L_o}{\partial B^i_a}\,\pi_a\ ,$$

$$(3.18.9)\quad \underset{\sim}{\mathcal{D}} = d\underset{\sim}{\phi}\ ,\qquad R_i = \frac{\partial L}{\partial \mathcal{D}^i_{ab}}\,\pi_{ab}\ ,$$

$$(3.18.10)\quad d\underset{\sim}{Z} = \underset{\sim}{0}\ ,$$

$$(3.18.11)\quad 2\,d\underset{\sim}{R} = \hat{\underset{\sim}{Z}} = \underset{\sim}{Z} + \underset{\sim}{Q}\ ,$$

and the boundary conditions

$$(3.18.12)\quad (\sigma_i^A - \partial_B Q^{BA} - \partial_4 Q^{4A})\Big|_{\partial B} N_A = 0\ ,$$

$$(3.18.13)\quad R_i^{aA}\Big|_{\partial B} N_A = 0\ .$$

(Note that $d\hat{Z}_i = dZ_i + d^2Q_i = dZ_i$ so that the balance of linear momentum equations, $dZ_i = 0$ are left invariant.)

Thus, the only changes are in the field equations (3.18.10) and in the boundary conditions (3.18.11).

Let us set

(3.18.14) $S_i^A = \partial_B Q_i^{BA} + \partial_4 Q_i^{4A}$, $P_i = -\partial_B Q_i^{B4}$.

Since $Q_i^{ab} + Q_i^{ba} = 0$, we then have

(3.18.15) $\partial_A S_i^A \equiv \partial_4 P_i$

while (3.18.11) and (3.18.12) give

(3.18.16) $s_1 \delta_{ji} \delta^{BD} [\partial^A(\partial_A \phi_D^i - \partial_D \phi_A^i) - \frac{1}{y} \partial_4(\partial_4 \phi_D^i - \partial_D \phi_4^i)]$

$$= \frac{1}{2}(\sigma_j^B - S_j^B) ,$$

(3.18.17) $\frac{s_1}{y} \delta_{ij} \partial^A(\partial_A \phi_4^i - \partial_4 \phi_A^i) = \frac{1}{2}(p_i - P_i)$,

(3.18.18) $\sigma_i^A \Big|_{\partial B} N_A = S_i^A \Big|_{\partial B} N_A = T_i(X^E) \Big|_\partial$,

$$p_i \Big|_{T=T_o} = P_i \Big|_{T=T_o} = \overset{o}{p}_i(X^A) ,$$

where $T_i(X^E)\Big|_{\partial B}$ are the specified surface tractions (see (3.15.7)). In view of (3.18.15) and (3.18.18) we may interpret S_i^A as the Piola-Kirchhoff stress and P_i

as the momentum that would be associated with an elasticity problem with the same initial and boundary data. Thus, $\sigma_i^A - S_i^A$ can be interpreted as the *effective Piola-Kirchhoff stress* and $p_i - P_i$ as the *effective linear momentum*. The field equations (3.18.17) and (3.18.18) then show that *the dislocation fields of finite bodies are driven by the effective stress and momentum*. This, however, is exactly what is known from detailed metallurgical studies: defects are driven by effective stress rather than the true stress σ_i^A in static and quasistatic processes. The only real difference here is that we have accounted for what happens in the dynamic case as well.

CHAPTER 4

LINEARIZATIONS

4.1 Group Scaling Parameters

The field equations (3.14.1-3) are a system of coupled nonlinear differential equations that reveal certain specific and useful insights, and yet are very difficult to solve. Clearly, a more complete understanding of the phenomena described by these field equations would be aided by some form of simplification.

Up to this point we have not said anything about units of measure for the compensating fields W^{α}_{a} and ϕ^{i}_{a} . The breaking of the homogeneity of the underlying group $G_o = SO(3)_o \triangleright T(3)_o$ gives rise to these disclination and dislocation field variables irrespective of coordinates in the group space. We therefore have the liberty to fix the units with respect to which defects will be measured. Since ϕ^i and W^{α} fields are naturally associated with the gauge group, a homogeneous scaling of the gauge group generators provides the means for achieving the required calibration.

Consider the connection $\hat{\underset{\sim}{\Gamma}}$ associated with the full group G

$$\hat{\underset{\sim}{\Gamma}} = \begin{pmatrix} W^{\alpha}\underset{\sim}{\gamma}_{\alpha} & \phi^{i}\underset{\sim}{t}_{i} \\ [\underset{\sim}{0}] & 0 \end{pmatrix} ,$$

and let ε be a group scaling parameter. The scaling $(\underset{\sim}{\gamma}_{\alpha}, \underset{\sim}{t}_{i}) \rightarrow (\varepsilon\underset{\sim}{\gamma}_{\alpha}, \varepsilon\underset{\sim}{t}_{i})$ of the gauge group generators induces the scaling

$$\hat{\underset{\sim}{\Gamma}} \rightarrow \varepsilon\,\hat{\underset{\sim}{\Gamma}} = \begin{pmatrix} \varepsilon\, W^{\alpha}\underset{\sim}{\gamma}_{\alpha} & \varepsilon\,\phi^{i}\underset{\sim}{t}_{i} \\ [\underset{\sim}{0}] & 0 \end{pmatrix} , \tag{4.1.1}$$

of the connection matrix. Evidently one could view this process as if the group generators $\underset{\sim}{\gamma}_{\alpha}$ and $\underset{\sim}{t}_{i}$ remain unchanged and instead, the compensating fields W^{α} and ϕ^{i} are scaled by the parameter ε .

$$W^{\alpha} \rightarrow \varepsilon\, W^{\alpha} \,,\; \phi^{i} \rightarrow \varepsilon\,\phi^{i} \;. \tag{4.1.2}$$

This natural scaling of the gauge group is used to introduce an expansion parameter, ε , that results in approximation procedures that are, in a sense, uniform. From (4.1.2) it follows that the parameter ε determines scale units of disclinations and dislocations.

The generating matrices $\underset{\sim}{\gamma}_{\alpha}$ of the rotation group SO(3) satisfy the relations

$$[\underset{\sim}{\gamma}_{\alpha}, \underset{\sim}{\gamma}_{\beta}] = C^{\delta}_{\alpha\beta}\,\underset{\sim}{\gamma}_{\delta} ,$$

where $C^{\delta}_{\alpha\beta}$ are the structure constants of the Lie group SO(3) . Hence, the scaling of the gauge group generators induces the scaling

$$C^{\alpha}_{\beta\gamma} \longrightarrow \varepsilon\, C^{\alpha}_{\beta\gamma}\ , \tag{4.1.3}$$

of the structure constants of the subgroup SO(3) . We note a similar situation in Yang-Mills theory [7], where the structure constants are also multiplied by a scaling constant that may be chosen at liberty.

The components of the non-singular Cartan-Killing metric $C_{\alpha\beta}$ of the subgroup SO(3) are given in terms of the structure constants of SO(3) by

$$C_{\alpha\beta} = C^{\gamma}_{\alpha\delta}\, C^{\delta}_{\beta\gamma}\ .$$

Thus, (4.1.3) implies

$$C_{\alpha\beta} \longrightarrow \varepsilon^{2}\, C_{\alpha\beta}\ , \tag{4.1.4}$$

and consequently

$$C^{\alpha\beta} \longrightarrow \frac{1}{\varepsilon^{2}}\, C^{\alpha\beta}\ . \tag{4.1.5}$$

We can now proceed with expansions in terms of the scaling parameter ε . Such expansions have a natural relation to expansions about the identity element in the gauge group space in view of the scaling laws

$$(\underset{\sim}{\gamma}_\alpha,\ \underset{\sim}{t}_i) \rightarrow (\varepsilon\underset{\sim}{\gamma}_\alpha,\ \varepsilon\underset{\sim}{t}_i)\ .$$

4.2 Approximate Equations - The ε Expansion

With the scaling parameter ε, the components of the distortion-velocity 1-forms can be written as

$$B^i_a = \partial_a \chi^i + \varepsilon \phi^i_a + \varepsilon \gamma^i_{\alpha j} W^\alpha_a \chi^j . \tag{4.2.1}$$

Let us also introduce the displacement vector $\{u^i\}$ by

$$\chi^i(X^b) = \delta^i_a X^a + u^i(X^b) . \tag{4.2.2}$$

Now that we have a specific parameter we proceed in the usual manner by expanding each of the functions u^i, ϕ^i_a, W^α_a in series in ascending powers of ε to give

$$\begin{cases} u^i = \underset{0}{u}^i + \varepsilon \underset{1}{u}^i + \varepsilon^2 \underset{2}{u}^i + \dots , \\ \phi^i_a = \underset{0}{\phi}^i_a + \varepsilon \underset{1}{\phi}^i_a + \varepsilon^2 \underset{2}{\phi}^i_a + \dots , \\ W^\alpha_a = \underset{0}{W}^\alpha_a + \varepsilon \underset{1}{W}^\alpha_a + \varepsilon^2 \underset{2}{W}^\alpha_a + \dots . \end{cases} \tag{4.2.3}$$

It is a lengthy, but simple calculation to write the field equations (3.14.1-3) and the constitutive relations (3.4. 6-9) in corresponding ε-expanded forms.

The four free parameters, y, ζ, s_1, s_2, of the theory play an obviously important role in understanding defect phenomena. The constants y and ζ are dynamic parameters; they are related to the speeds of propagation of the dislocation and disclination fronts, respectively. The parameters s_1 and s_2 are important in both the static and the dynamic cases. They may be thought of as characteristic of the energies required in order to create "unit" dislocation and "unit" disclination, respectively. The ratios of the elastic energy to the dislocation energy and of the dislocation energy to the disclination energy characterized by the ratios λ/s_1 and s_1/s_2, respectively are, in fact, the distinguishing parameters of the theory. For our purposes, let λ/s_1 and s_1/s_2 each be of order ε ; that is

$$(4.2.4) \qquad \frac{\lambda}{s_1} = \varepsilon\widetilde{\left(\frac{\lambda}{s_1}\right)} \equiv \varepsilon L_1 \ , \quad \frac{\lambda}{s_2} = \varepsilon^2\widetilde{\left(\frac{\lambda}{s_2}\right)} \equiv \varepsilon^2 L_2$$

in the approximation considered here. Other choices of the ε-orders of λ/s_1 and s_1/s_2 will result in the emergence of different effects. This is as one would expect; different ε-orders of λ/s_1 and s_1/s_2 lead to models of different physical phenomena.

We take the order of the model to be that determined by the power of ε in the expression for the basic variables B^i_a given by

$$(4.2.5)\qquad B^i_a = \delta^i_a + \partial_a \underset{o}{u}^i + \varepsilon[\partial_a \underset{1}{u}^i + \underset{o}{\phi}^i_a + \gamma^i_{\alpha j} \underset{o}{W}^\alpha_a(\delta^j_A X^A + \underset{o}{u}^j)] + \varepsilon^2\{\partial_a \underset{2}{u}^i + \underset{1}{\phi}^i_a + \gamma^i_{\alpha j}[\underset{o}{W}^\alpha_a \underset{1}{u}^j + \underset{1}{W}^\alpha_a(\delta^j_B X^B + \underset{o}{u}^j)]\} + \ldots .$$

Hence, in the first order approximation, the phenomena are described by the functions $\underset{1}{u}^i$, $\underset{o}{\phi}^i_a$ and $\underset{o}{W}^\alpha_a$.

The requirement that the current and reference configurations coincide at time $T = 0$ in the lowest order implies

$$(4.2.6)\qquad \underset{o}{u}^i = 0 .$$

Moreover, since $\underset{o}{\phi}^i_a$, $\underset{o}{W}^\alpha_a$ and $\underset{1}{W}^\alpha_a$ satisfy homogeneous partial differential equations (for the choice (4.2.4)) with homogeneous initial data (recall that the reference state is defect free), we obtain only the trivial solutions,

$$(4.2.7)\qquad \underset{o}{\phi}^i_a = 0 , \quad \underset{o}{W}^\alpha_a = \underset{1}{W}^\alpha_a = 0 .$$

Then by (4.2.6-7), the expressions (4.2.5) become

$$(4.2.8)\qquad B^i_a = \delta^i_a + \varepsilon\, \partial_a \underset{1}{u}^i + \varepsilon^2(\partial_a \underset{2}{u}^i + \underset{1}{\phi}^i_a) + \varepsilon^3(\partial_a \underset{3}{u}^i + \underset{2}{\phi}^i_a + \gamma^i_{\alpha j} \underset{2}{W}^\alpha_a \delta^j_B X^B) + \ldots .$$

We point out that the approximation procedure involved here is not the customary engineering linearization. We establish the approximations based solely on the homogeneous scaling of the group parameters that, in turn, induced the ε-expansion of the field variables.

In the first order approximation, the only field equations that need to be satisfied are the equations of balance of linear momentum,

$$(4.2.9) \qquad \partial_4 \underset{1}{p}_i - \partial_A \underset{1}{\sigma}_i^A = 0 ,$$

with the constitutive relations

$$(4.2.10) \qquad \begin{cases} \underset{1}{\sigma}_i^A = \frac{\lambda}{2} (\delta_i^A \delta^{FD} \underset{1}{E}_{FD} + \frac{2\mu}{\lambda} \delta_i^S \delta^{RA} \underset{1}{E}_{RS}) , \\ \underset{1}{E}_{AB} = \delta_{Bj} \partial_A \underset{1}{u}^j + \delta_{Aj} \partial_B \underset{1}{u}^j , \\ \underset{1}{p}_i = \rho_o \delta_{ij} \partial_4 \underset{1}{u}^j . \end{cases}$$

It is, therefore, evident that in this lowest order we recover elasticity theory. The field variables ϕ_a^i and W_a^α that describe dislocations and disclinations do not occur in the first order approximation.

The elastic stresses, $\underset{1}{\sigma}_i^A$, and momenta, $\underset{1}{p}_i$, serve as sources in the balance of dislocation equations in the second order approximation. The field equations of defect dynamics in the second order approximation are

$$(4.2.11)\quad \begin{cases} \partial_4 \underset{2}{p}_i - \partial_A \underset{2}{\sigma}_i^A = 0 , \\ \delta_{ji}\delta^{BD}[\partial^A \partial_A \underset{1}{\phi}_D^i - \partial_D\partial^A \underset{1}{\phi}_A^i - \frac{1}{y}\partial_4(\partial_4 \underset{1}{\phi}_D^i - \partial_D \underset{1}{\phi}_4^i)] \\ \qquad = \frac{1}{2} L_1 \underset{1}{\hat{\sigma}}_j^B , \\ \frac{1}{y}\delta_{ij}\partial^A(\partial_A \underset{1}{\phi}_4^i - \partial_4 \underset{1}{\phi}_A^i) = \frac{1}{2\lambda} L_1 \underset{1}{p}_j , \end{cases}$$

and the constitutive relations are given by

$$(4.2.12)\quad \begin{cases} \underset{2}{\hat{\sigma}}_i^A = \frac{1}{2}\delta_{ij}\{\delta^{Aj}\delta^{FD} \underset{2}{E}_{FD} + \frac{2\mu}{\lambda}\delta^{RA}\delta^{Sj} \underset{2}{E}_{RS} \\ \qquad + \partial_C \underset{1}{u}^j(\delta^{AC}\delta^{FD} \underset{1}{E}_{FD} + \frac{2\mu}{\lambda}\delta^{RA}\delta^{SC} \underset{1}{E}_{RS})\} , \\ \underset{2}{E}_{AB} = \delta_{Aj}(\partial_B \underset{2}{u}^j + \underset{1}{\phi}_B^j) + \partial_A \underset{1}{u}^i \delta_{ij} \partial_B \underset{1}{u}^j \\ \qquad + \delta_{Bi}(\partial_A \underset{2}{u}^i + \underset{1}{\phi}_A^i) , \\ \underset{2}{p}_i = \rho_o\delta_{ij}(\partial_4 \underset{2}{u}^j + \underset{1}{\phi}_4^j) . \end{cases}$$

Here, we have set

$$(4.2.13)\quad \hat{\sigma}_i^A = \frac{1}{\lambda}\sigma_i^A ,$$

and similarly,

$$(4.2.14)\quad \hat{j}_\alpha^a = \frac{1}{\lambda} j_\alpha^a .$$

The stresses, $\underset{2}{\sigma}{}_i^A$, and momenta, $\underset{2}{p}{}_i$, are not elastic in the second order approximation. They take into account the presence of dislocation fields. The terms $\partial_C \underset{1}{u}{}^j (\delta^{AC} \delta^{FD} \underset{1}{E}{}_{FD} + 2\mu/\lambda\, \delta^{RA} \delta^{SC} \underset{1}{E}{}_{RS})$ in the expression for $\underset{2}{\sigma}{}_i^A$ come from the multiplication of $\partial L_o/\partial E_{RS}$ by B_F^k and $B_F^k = \delta_F^k + \varepsilon \partial_F \underset{1}{u}{}^k$ in this order of approximation and render $\underset{2}{\sigma}{}_i^A$ unsymmetric in the indices (i,A) . Similarly, the terms $\partial_A \underset{1}{u}{}^i \delta_{ij} \partial_B \underset{1}{u}{}^j$ in $\underset{2}{E}{}_{AB}$ come from the full non-linear strain measure $B_A^i \delta_{ij} B_B^j - \delta_{AB}$. Both of these sets of terms are absent in the usual engineering approximation. The first set of equations in (4.2.11) are second order refinements of the statements of the balance of linear momentum in the theory of dislocation dynamics. The second and third sets in (4.2.11) serve for the determination of the field variables $\underset{1}{\phi}{}_a^i$. The right-hand sides of those equations are known from the first order approximation. We may thus say that the second order approximation models dislocation dynamics.

Only in the third and higher orders approximation do the compensating fields W_a^α (disclination phenomena) enter the equations. The third order approximation field equations assume the following form

$$(4.2.15) \qquad \partial_4 \underset{3}{p}{}_i - \partial_A \underset{3}{\sigma}{}_i^A = 0 \ ;$$

$$(4.2.16)\quad \begin{cases} \partial^A \underset{2}{\bar{\mathcal{D}}}{}^i_{AD} - \frac{1}{y}\,\partial_4 \underset{2}{\bar{\mathcal{D}}}{}^i_{4D} + \gamma^i_{\alpha k}[\partial^A(\underset{2}{F}{}^\alpha_{AD}\,\delta^k_C\,X^C) \\ \qquad - \frac{1}{y}\,\partial_4(\underset{2}{F}{}^\alpha_{4D}\,\delta^k_C\,X^C)] = \frac{1}{2}\,L_1\,\delta^{ij}\,\delta_{DB}\,\underset{2}{\hat{\sigma}}{}^B_j\ , \\ \frac{1}{y}[\partial^A \underset{2}{\bar{\mathcal{D}}}{}^i_{A4} + \gamma^i_{\alpha k}\,\partial^A(\underset{2}{F}{}^\alpha_{A4}\,\delta^k_C\,X^C)] = \frac{1}{2\lambda}\,L_1\delta^{ij}\,\underset{2}{p}_j\ ; \end{cases}$$

$$(4.2.17)\quad \begin{cases} \partial^A(\partial_A \underset{2}{W}{}^\rho_C - \partial_C \underset{2}{W}{}^\rho_A) - \frac{1}{\zeta}\,\partial_4(\partial_4 \underset{2}{W}{}^\rho_C - \partial_C \underset{2}{W}{}^\rho_4) \\ \qquad = -\frac{1}{2}\,L_2\delta_{CB}\,C^{\rho\eta}\,\underset{2}{\hat{\jmath}}{}^B_\eta \\ \frac{1}{\zeta}\,\partial^A(\partial_A \underset{2}{W}{}^\rho_4 - \partial_4 \underset{2}{W}{}^\rho_A) = \frac{1}{2}\,L_2 C^{\rho\eta}\,\underset{2}{\hat{\jmath}}{}^4_\eta\ . \end{cases}$$

Here,

$$\bar{\mathcal{D}}^i_{ab} = \varepsilon^2\,\underset{2}{\bar{\mathcal{D}}}{}^i_{ab} + \varepsilon^3\,\underset{3}{\bar{\mathcal{D}}}{}^i_{ab} + \ldots\ ,$$

$$F^\alpha_{ab} = \varepsilon^2\,\underset{2}{F}{}^\alpha_{ab} + \varepsilon^3\,\underset{3}{F}{}^\alpha_{ab} + \ldots\ .$$

The constitutive relations are

$$(4.2.18a)\quad \underset{3}{\sigma}{}^A_i = \frac{1}{2}\,\delta_{ij}\{\delta^{Aj}\,\delta^{FD}\,\underset{3}{E}_{FD} + \frac{2\mu}{\lambda}\,\delta^{RA}\,\delta^{Sj}\,\underset{3}{E}_{RS}$$
$$+ \partial_C\,\underset{1}{u}{}^j(\delta^{AC}\,\delta^{FD}\,\underset{2}{E}_{FD} + \frac{2\mu}{\lambda}\,\delta^{RA}\,\delta^{SC}\,\underset{2}{E}_{RS})$$
$$+ (\partial_C\,\underset{2}{u}{}^j + \underset{1}{\phi}{}^j_C)(\delta^{AC}\,\delta^{FD}\,\underset{1}{E}_{FD}$$
$$+ \frac{2\mu}{\lambda}\,\delta^{RA}\,\delta^{SC}\,\underset{1}{E}_{RS})\}\ ,$$

(4.2.18b) $$\underset{3}{E}_{AB} = \delta_{Aj}(\partial_B \underset{3}{u}^j + \underset{2}{\phi}^j_3 + \gamma^j_{\alpha k} \underset{2}{W}^\alpha_B \delta^k_C X^C)$$

$$+ \partial_A \underset{1}{u}^i \delta_{ij}(\partial_B \underset{2}{u}^j + \underset{1}{\phi}^j_B) + (\partial_A \underset{2}{u}^i$$

$$+ \underset{1}{\phi}^i_A)\delta_{ij} \partial_B \underset{1}{u}^j$$

$$+ (\partial_A \underset{3}{u}^i + \underset{2}{\phi}^i_A + \gamma^i_{\alpha k} \underset{2}{W}^\alpha_A \delta^k_C X^C)\delta_{Bi} ,$$

$$\underset{3}{p}_i = \rho_o \delta_{ij}(\partial_4 \underset{3}{u}^j + \underset{2}{\phi}^j_4 + \gamma^j_{\alpha k} \underset{2}{W}^\alpha_4 \delta^k_B X^B) ,$$

(4.2.19) $$\begin{cases} \underset{2}{\hat{j}}^A_\alpha = - \dfrac{2}{L_1} \gamma^i_{\alpha j} \delta_{ik} \delta^{jC} \delta^{AD} \underset{1}{\bar{D}}^k_{CD} , \\ \underset{2}{\hat{j}}^4_\alpha = - \dfrac{2}{yL_1} \gamma^i_{\alpha j} \delta_{ik} \delta^{jC} \underset{1}{\bar{D}}^k_{4C} . \end{cases}$$

From (4.2.17) and (4.2.19) we see that disclinations are driven by dislocations in the third order approximation. The balance of disclination equations (4.2.17) are linear second order partial differential equations. We note that the balance of linear momentum equations (4.2.15) with disclination fields present in the third order look exactly the same as in lower orders approximations. The forces due to defects do not enter until we consider at least the fourth order approximation.

The third order approximation models defect dynamics,

with both dislocations and disclinations present. Maybe the fact that disclinations do not enter the field equations until at least the third order approximation is the essential conceptual obstacle that accounts for their exclusion in most theories of defects. We recall that this approximation procedure was based on the assumption that the required disclination energy is very large compared with the dislocation energy and this, in turn, is very large compared with the elastic energy (i.e. $\lambda/s_1 \sim \varepsilon$, $s_1/s_2 \sim \varepsilon$) . Without this assumption we would get significantly different models.

The balance of moment of momentum equations in this approximation procedure occur for the first time in the second order: $\gamma^i_{\alpha j} \underset{1}{\sigma}{}^A_i \delta^j_A = 0$.
They simply state that the elastic stress tensor is symmetric. However, in the third order approximation the $\underset{\sim}{\phi}$ and $\underset{\sim}{\chi}$ variables have to satisfy the following conditions

$$(4.2.20) \qquad \gamma^i_{\alpha j}(\underset{2}{\sigma}{}^A_i \delta^j_A + \underset{1}{\sigma}{}^A_i \partial_A \underset{1}{u}{}^j) = 0$$

Higher order approximations do not reveal essentially new informations. On the other hand the equations themselves and the constitutive relations become algebraically significantly more complex.

The governing equations become significantly simpler in the static case. They, in fact, reduce to the familiar equations of magnetostatics. To show this, let us define vectors $\vec{\Sigma}^{i}$, $\vec{\Phi}^{i}$, $\vec{M}^{\alpha}$ and $\vec{P}^{\alpha}$ by

$$(4.2.21)\quad \begin{cases} \vec{\Phi}^{i} = \phi^{i}_{1}\,\vec{e}_{1} + \phi^{i}_{2}\,\vec{e}_{2} + \phi^{i}_{3}\,\vec{e}_{3}\,, \\ \vec{\Sigma}^{i} = \delta^{ij}(\hat{\sigma}^{1}_{j}\,\vec{e}_{1} + \hat{\sigma}^{2}_{j}\,\vec{e}_{2} + \hat{\sigma}^{3}_{j}\,\vec{e}_{3})\,, \\ \vec{M}^{\alpha} = W^{\alpha}_{1}\,\vec{e}_{1} + W^{\alpha}_{2}\,\vec{e}_{2} + W^{\alpha}_{3}\,\vec{e}_{3}\,, \\ \vec{P}^{\alpha} = C^{\alpha\beta}(\hat{J}^{1}_{\beta}\,\vec{e}_{1} + \hat{J}^{2}_{\beta}\,\vec{e}_{2} + \hat{J}^{3}_{\beta}\,\vec{e}_{3})\,. \end{cases}$$

Here, the set of orthonormal vectors $\{\vec{e}_{A}\}$ spans the 3-dimensional space E_3 with coordinate cover $\{X^{A}\}$. With the expressions (4.2.21), the second order approximation equations (4.2.11) can be rewritten in the following way

$$(4.2.22)\quad \vec{\nabla} \times \vec{\nabla} \times \underset{1}{\vec{\Phi}^{i}} = -\frac{1}{2} L_{1} \underset{1}{\vec{\Sigma}^{i}}\,.$$

Similarly, in the third order approximation, from (4.2.16-17) we obtain static equations for dislocations

$$(4.2.23)\quad \vec{\nabla} \times \vec{\nabla} \times \underset{2}{\vec{\Phi}^{i}} = -\frac{1}{2}(L_{1} \underset{2}{\vec{\Sigma}^{i}} + L_{2}\gamma^{i}_{\alpha k}\,\delta^{k}_{C}\,X^{C}\,\underset{2}{\vec{P}^{\alpha}}) - \vec{a}^{i}_{\alpha} \times \vec{\nabla} \times \underset{2}{\vec{M}^{\alpha}}\,,$$

$\vec{a}^{i}_{\alpha} = \delta^{kA}\,\gamma^{i}_{\alpha k}\,\vec{e}_{A}$, and for disclinations

$$(4.2.24)\quad \vec{\nabla} \times \vec{\nabla} \times \underset{2}{\vec{M}^{\alpha}} = \frac{1}{2} L_{2} \underset{2}{\vec{P}^{\alpha}}\,.$$

In the higher order approximations the governing equations for dislocations and disclinations are of the same kind as (4.2.23-24).

Since $\vec{\nabla}\cdot(\vec{\nabla} \times \vec{\nabla} \times \vec{\Psi}) \equiv 0$, it follows that the divergences of the vectors on the right-hand sides of (4.2.23-24) have to vanish. Similar situations prevail in the higher order approximations. These conditions will either be identically satisfied as a consequence of the field equations and the statements of balance in the previous orders of approximation, or they represent further restrictions on the field variables. It might seem that we are imposing consistency conditions that do not occur in the original theory. However, this is not the case. The integrability conditions for the field equations for dislocations and disclinations gave the equations of the balance of linear momentum and moment of momentum, respectively. In the approximation procedure, since the equations that occur are essentially nonlinear, it is not true that the integrability conditions are preserved. Therefore, at each order of the approximation they have to be secured anew.

In the second and third order approximations, the integrability conditions for the equations (4.2.22-24), are identically satisfied as a consequence of the field equations and the balance equations. Therefore, in the third order approximation it remains to satisfy equations (4.2.20) of the balance of moment of momentum, while in the second order approximation the only requirement is that the Piola-Kirchhoff elastic stress tensor be symmetric. In the absence of disclinations, the integrability condition for (4.2.23) are always satisfied; they are equivalent to equations (4.2.11) that have already been solved in the previous order approximation.

We have mentioned that different choices of the ε-order of the ratios λ/s_1 and s_1/s_2 lead to models of different phenomena. For the particular choice made in this section, elasticity theory is recovered in the first order approximation. The second order models materials with dislocations, while disclinations do not occur until at least the third order approximation. We do not intend to give a detailed analysis of those different choices of the ε-order of λ/s_1 and s_1/s_2, but some comments seem to be appropriate.

For a moment, suppose that the parameters λ and s_1 (coefficients of elastic and dislocation energy densities, respectively) are commensurate and that s_2 is large compared with s_1. As one would expect in this case, the dislocation field variables ϕ^i_a and the displacement functions u^i enter the field equations in the same (lowest) order approximation. This means that even in the lowest order in the ε-expansion we do not recover pure elasticity. The stresses that drive the dislocation equations are dislocation dependent. Disclination effects become noticible in the next higher order approximation.

The theory is open to different possibilities. Depending on the choice of the coupling constants different physical phenomena are modeled. But, once the coupling constants are fixed, the model for a particular physical phenomena is uniquely determined.

4.3 Homogeneous Deformation

We will now illustrate how the presence of the dislocation fields ϕ^i_A effects the displacements u^i in states of static homogeneous deformations. Since we are not interested in disclinations it means that the first order approximation of the previous section should model the elastic material properties that we want to consider (i.e., first order stresses balance any and all applied tractions on any boundary). This, in turn, implies that the second order stresses $\underset{2}{\sigma}{}^A_i$ have to vanish. Hence, the governing equations for the static homogeneous deformation of the materials with dislocations are

$$(4.3.1) \qquad \partial^A \partial_A \underset{1}{\phi}{}^i_D - \partial_D \partial^A \underset{1}{\phi}{}^i_A = \frac{1}{2} \delta^{ij} \delta_{DB} L_1 \underset{1}{\hat{\sigma}}{}^B_j ,$$

and the constitutive relations are given by

$$(4.3.2) \quad \begin{cases} \underset{2}{\hat{\sigma}}{}^A_i = \frac{1}{2} \delta_{ij} \{ \delta^{Aj} \operatorname{tr} \underset{2}{\tilde{E}} + \frac{2\mu}{\lambda} \delta^{RA} \delta^{Sj} \underset{1}{E}_{RS} \\ \qquad + \partial_C \underset{1}{u}{}^j (\delta^{AC} \operatorname{tr} \underset{1}{\tilde{E}} + \frac{2\mu}{\lambda} \delta^{RA} \delta^{SC} \underset{1}{E}_{RS}) \} = 0 , \\ \underset{2}{E}_{AB} = \delta_{Aj} (\partial_B \underset{2}{u}{}^j + \underset{1}{\phi}{}^j_B) + \partial_A \underset{1}{u}{}^i \delta_{ij} \partial_B \underset{1}{u}{}^j \\ \qquad + \delta_{Bj} (\partial_A \underset{2}{u}{}^j + \underset{1}{\phi}{}^j_A) . \end{cases}$$

The stress and strain tensors, $\underset{1}{\tilde{\hat{\sigma}}}$ and $\underset{1}{\tilde{E}}$, respectively, are given by the standard relations of linear elasticity theory.

The field equations (4.3.1) are gauge invariant, and solutions are determined to within a gauge. Therefore, we impose the Lorentz gauge

$$(4.3.3) \qquad \partial^A \underset{1}{\phi}{}^i_A = 0 \ .$$

The equations (4.3.1) then become

$$(4.3.4) \qquad \nabla^2 \underset{1}{\phi}{}^i_D = \frac{1}{2} \delta^{ij} \delta_{BD} L_1 \underset{1}{\hat{\sigma}}{}^B_j \ .$$

Since we did not impose antiexact gauge conditions we remind the reader that the displacement fields $\underset{2}{u}{}^i$ are not the total displacement fields for the second order corrections.

We proceed by an explicit example. Consider simple shear of a rectangular block with the plane of shear being the XY-plane. The displacement gradient matrix for such configurations is given by

$$(4.3.5) \qquad ((\partial_A \underset{1}{u}{}^i)) = \begin{pmatrix} 0 & \frac{a}{2} & 0 \\ \frac{a}{2} & 0 & 0 \\ 0 & 0 & 0 \end{pmatrix}$$

From (4.2.9-10), we obtain the elastic strain and stress

$$(4.3.6) \qquad \underset{1}{\tilde{E}} = \begin{pmatrix} 0 & a & 0 \\ a & 0 & 0 \\ 0 & 0 & 0 \end{pmatrix} \quad \text{and} \quad \underset{1}{\tilde{\sigma}} = \begin{pmatrix} 0 & \mu a & 0 \\ \mu a & 0 & 0 \\ 0 & 0 & 0 \end{pmatrix} \ .$$

The elastic stress $\underset{1}{\sigma}$ is the driving term for the dislocations in the second order approximation. From (4.3.4) it follows that $\underset{1}{\phi}{}^{1}_{2}$ and $\underset{1}{\phi}{}^{2}_{1}$ satisfy Poisson's equations

$$(4.3.7) \qquad \begin{aligned} \nabla^2 \, \underset{1}{\phi}{}^{1}_{2} &= \frac{1}{2} L_1 \frac{\mu}{\lambda} a , \\ \nabla^2 \, \underset{1}{\phi}{}^{2}_{1} &= \frac{1}{2} L_1 \frac{\mu}{\lambda} a , \end{aligned}$$

while the other ϕ's are solutions of Laplace's equation.

For our purposes, we need to find a solution for the fields ϕ^{i}_{a} . Let

$$(4.3.8) \qquad \delta = \frac{1}{2} L_1 \frac{\mu}{\lambda} a .$$

It suffices to take trivial solutions for the Laplace's equations, that is

$$(4.3.9) \qquad \underset{1}{\phi}{}^{1}_{1} = \underset{1}{\phi}{}^{2}_{2} = \underset{1}{\phi}{}^{3}_{3} = \underset{1}{\phi}{}^{2}_{3} = \underset{1}{\phi}{}^{1}_{3} = \underset{1}{\phi}{}^{3}_{1} = \underset{1}{\phi}{}^{3}_{2} = 0 .$$

The only nonzero entries in the matrix $\underset{1}{\phi}$ are then $\underset{1}{\phi}{}^{1}_{2}$ and $\underset{1}{\phi}{}^{2}_{1}$ and they satisfy equations (4.3.7) and the Lorentz gauge (4.3.3). If we take

$$(4.3.10) \qquad \underset{1}{\phi}{}^{1}_{2} = \frac{}{2} X^2 \quad \text{and} \quad \underset{1}{\phi}{}^{2}_{1} = \frac{\delta}{2} Y^2 ,$$

then equations (4.3.7) and conditions (4.3.3) are satisfied. From (4.3.2) we obtain the components of the strain tensor in the second order approximation

$$
(4.3.11)\quad
\begin{cases}
\underset{2}{E}_{11} = \dfrac{a^2}{4} + 2\partial_1 \underset{2}{u}^1 , \\
\underset{2}{E}_{22} = \dfrac{a^2}{4} + 2\partial_2 \underset{2}{u}^2 , \\
\underset{2}{E}_{33} = 0 , \\
\underset{2}{E}_{12} = \underset{2}{E}_{21} = \partial_2 \underset{2}{u}^1 + \partial_1 \underset{2}{u}^2 + \dfrac{\delta}{2}(X^2 + Y^2) , \\
\underset{2}{E}_{13} = \underset{2}{E}_{31} = \partial_3 \underset{2}{u}^1 , \\
\underset{2}{E}_{23} = \underset{2}{E}_{32} = \partial_3 \underset{2}{u}^2 ,
\end{cases}
$$

The conditions $\underset{2}{\sigma}^A_i = 0$ thus give differential equations for the displacements $\underset{2}{u}^i$:

$$
(4.3.12)\quad
\begin{cases}
\dfrac{a^2}{2} + 2(\partial_1 \underset{2}{u}^1 + \partial_2 \underset{2}{u}^2) + 2\,\dfrac{\mu}{\lambda}\,(\dfrac{a^2}{4} + 2\partial_1 \underset{2}{u}^1 + \dfrac{a^2}{2}) \\
\qquad = 0 , \\
\dfrac{a^2}{2} + 2(\partial_1 \underset{2}{u}^1 + \partial_2 \underset{2}{u}^2) + 2\,\dfrac{\mu}{\lambda}\,(\dfrac{a^2}{4} + 2\partial_2 \underset{2}{u}^2 + \dfrac{a^2}{2}) \\
\qquad = 0 , \\
\dfrac{a^2}{2} + 2(\partial_1 \underset{2}{u}^1 + \partial_2 \underset{2}{u}^2) = 0 , \\
\partial_2 \underset{2}{u}^1 + \partial_1 \underset{2}{u}^2 + \dfrac{\delta}{2}\,(X^2 + Y^2) = 0 , \\
\partial_3 \underset{2}{u}^1 = 0 , \\
\partial_3 \underset{2}{u}^2 = 0 .
\end{cases}
$$

A solution of the system (4.3.12) is given by

$$(4.3.13)\quad \begin{cases} \underset{2}{u}^1 = -\dfrac{3a^2}{8}X - \dfrac{\delta}{6}Y^3 \\[2ex] \underset{2}{u}^2 = -\dfrac{3a^2}{8}Y - \dfrac{\delta}{6}X^3 \\[2ex] \underset{2}{u}^3 = 0 . \end{cases}$$

We have noted that by choosing the Lorentz gauge the 1-forms ϕ^i are not antiexact. Therefore, the total integrable displacements $\underset{2}{u}^i_{tot}$ in the second order approximation are given by

$$(4.3.14)\qquad \underset{2}{u}^i_{tot} = \underset{2}{u}^i + H\underset{1}{\phi}^i .$$

From (4.3.10) it follows that

$$(4.3.15)\qquad H\underset{1}{\phi}^1 = \frac{\delta}{6}X^2Y \quad \text{and} \quad H\underset{1}{\phi}^2 = \frac{\delta}{6}XY^2$$

Finally, we can write the total integrable displacements for simple shear with dislocations present

$$u^i_{tot} = \varepsilon\underset{1}{u}^i + \varepsilon^2(\underset{2}{u}^i + H\underset{1}{\phi}^i) .$$

By (4.3.8) and (4.3.13-15), we obtain the explicit expressions

$$(4.3.16)\quad \begin{cases} u^1_{tot} = \frac{1}{2}\,\varepsilon a\{Y - \frac{3a}{4}\,\varepsilon X + \frac{L_1}{6}\,\frac{\mu}{\lambda}\,\varepsilon Y(X^2 - Y^2)\} \\ u^2_{tot} = \frac{1}{2}\,\varepsilon a\{X - \frac{3a}{4}\,\varepsilon Y + \frac{L_1}{6}\,\frac{\mu}{\lambda}\,\varepsilon X(Y^2 - X^2)\} \\ u^3_{tot} = 0\ . \end{cases}$$

The first and second terms on the right-hand sides of (4.3.16) are due to the pure elastic response, the first ones due to the linear elasticity and the second terms account for the non-linear elasticity effects. The third terms in the expressions for the displacements are due to the presence of the dislocations.

From (4.3.10) we can write the 1-forms $\underset{1}{\phi}^i$ and their exterior derivatives $d\underset{1}{\phi}^i$ as follows

$$(4.3.17)\quad \begin{aligned} \underset{1}{\phi}^1 &= \frac{\delta}{2}\,X^2 dY\ , & d\underset{1}{\phi}^1 &= \delta X\ dX \wedge dY \\ \underset{1}{\phi}^2 &= \frac{\delta}{2}\,Y^2 dX\ , & d\underset{1}{\phi}^2 &= -\ \delta Y\ dX \wedge dY \end{aligned}$$

Then, by (3.7.13), the Burgers vector is given by

$$(4.3.18)\quad \{b^i(\partial S_2)\} = \delta \int_{S_2} \begin{Bmatrix} X \\ -Y \\ 0 \end{Bmatrix} dX \wedge dY\ .$$

It lies in the XY-plane.

4.4 The Far Field of a Static Disclination - The Static Solution of Yang and Wu

The approximation procedure based upon expansion in the scaling parameter, ε , of the gauge group is effective because of the assumption that the reference state is defect free. This allows us to take zero Cauchy data for the ϕ^i_a and the W^α_a fields, in which case all solutions of the associated homogeneous equations are put equal to zero. There are, however, other possibilities, one of which we examine in this Section.

Let us restrict attention to phenomena where there is no breaking of the homogeneity of the action of the translation subgroup. The compensating fields ϕ^i_a are thus to be ignored from the beginning and the theory is governed by the disclination Lagrangian $L = L_o - s_2 L_2$. This is exactly the situation considered at the end of Section 3.15, where the governing field equations were shown to be

$$\text{(4.4.1)} \qquad D\bar{\underset{\sim}{G}} = \frac{1}{2}\bar{\underset{\sim}{J}} , \quad D\underset{\sim}{Z} = \underset{\sim}{0} \; . ,$$

with the constitutive relations

$$\text{(4.4.2)} \qquad \begin{aligned} 2\bar{\underset{\sim}{G}} &= - s_2 C^{\alpha\beta}(\partial L_2/\partial F^\alpha_{ab})\pi_{ab}\underset{\sim}{\gamma}_\beta , \\ Z_i &= (\partial L_o/\partial B^i_a)\pi_a , \quad \bar{\underset{\sim}{J}} = C^{\alpha\beta}(\underset{\sim}{Z}\underset{\sim}{\gamma}_\alpha\underset{\sim}{\chi})\underset{\sim}{\gamma}_\beta \; . \end{aligned}$$

If we now use the first of (4.4.2) to write $\bar{\underset{\sim}{G}} = s_2 \hat{\underset{\sim}{G}}$, then (4.4.1) gives

$$\text{(4.4.3)} \qquad D\hat{\underset{\sim}{G}} = \frac{1}{2s_2} \bar{\underset{\sim}{J}} = \frac{1}{2s_2} C^{\alpha\beta} (\underset{\sim}{Z} \underset{\sim}{\gamma}_\alpha \underset{\sim}{\chi}) \underset{\sim}{\gamma}_\beta$$

Let us restrict our attention to a region R of E_4 where all of the coefficients of the scalar-valued 3-forms $\underset{\sim}{Z}\underset{\sim}{\gamma}_\alpha \underset{\sim}{\chi}/2s_2$ are sufficiently small that (4.4.3) may be replaced by the approximate equation

$$\text{(4.4.4)} \qquad D\hat{\underset{\sim}{G}} \doteq \underset{\sim}{0} .$$

The field equations $D\hat{\underset{\sim}{G}} = \underset{\sim}{0}$ are, however, just the Yang-Mills free field equations since $SO(3)$ and $SU(2)$ are homeomorphic (recall that the connection matrix $\underset{\sim}{\Gamma}$ used to form the covariant exterior derivative D is a matrix of 1-forms that determines a connection on $SO(3)$). We may therefore make direct use of the known real-valued solutions of the Yang-Mills free field equations upon discarding the assumption that the reference configuration is defect free.

The simplest solution of the Yang-Mills free field equations is the static solution of Yang and Wu [16],

$$\text{(4.4.5)} \qquad W_A^\alpha = \delta^{\alpha\beta} \varepsilon_{\beta AB} X^B r^{-2} , \quad W_4^\alpha = 0 ,$$

with $r^2 = (X^1)^2 + (X^2)^2 + (X^3)^2$. This solution is already antiexact since $X^A W^\alpha_A + T W^\alpha_4 = 0$. When (4.4.5) is used to exhibit the 1-forms $W^\alpha = W^\alpha_a dX^a$, we have

(4.4.6) $$W^\alpha = \delta^{\alpha\beta} \varepsilon_{\beta AB} X^B r^{-2} dX^A .$$

Since the 1-form of the vector potential of a magnetostatic dipole with dipole moment m^i is $m^i \varepsilon_{iAB} X^B r^{-3} dX^A$, (4.4.6) shows that W^α is r multiplied by the field of a magnetostatic dipole with unit dipole moment in the α-direction. The multiplicative factor, r , may thus be viewed as the consequence of the nonlinearity of the Yang-Mills free field equations, and the three fields of 1-forms W^1 , W^2 , W^3 can be interpreted as a state whose character is similar to the presence of three magnetostatic dipoles at the origin of coordinates in E_3 with unit dipole moments coincident with the three spatial axes. Thus, the field of a static disclination in the region R looks like the composition of the fields of three magnetostatic dipoles at the origin with unit dipole moments coincident with the three spatial axes of coordinates and each multiplied by the factor r .

Now that we know the W-fields in the region R , we can go one step further and calculate the corresponding stress field. All that has to be done is to solve the field equations $\underset{\sim}{D}\underset{\sim}{Z} = \underset{\sim}{0}$; that is,

$$\partial_A \sigma_i^A = \gamma_{\alpha i}^{j} W_A^{\alpha} \sigma_j^A \tag{4.4.7}$$

with $p_i = 0$. If we put

$$\sigma_i^A = S_i^A g(r) , \tag{4.4.8}$$

the system (4.4.7) becomes

$$S_i^A \frac{dg}{dr} \partial_A r = - \gamma_{\alpha i}^{j} \varepsilon^{\alpha B \cdot}{}_{A} S_j^A g r^{-1} \partial_B r . \tag{4.4.9}$$

Now, (4.4.9) is satisfied by

$$g(r) = g_o r^m \tag{4.4.10}$$

where m and S_i^A are determined by the eigenvalue problem

$$- \gamma_{\alpha i}^{j} \varepsilon^{\alpha B \cdot}{}_{A} S_j^A = m S_i^B . \tag{4.4.11}$$

It is easily seen that (4.4.11) is satisfied by

$$m = -2 , \quad S_i^B = \delta_i^B , \tag{4.4.12}$$

in which case (4.4.8) gives

$$\sigma_i^A = g_o r^{-2} \delta_i^A . \tag{4.4.13}$$

The stress field distribution for the region R that arises from the disclination fields (4.4.5) has the coordinate axes as principal axes and dies out like r^{-2} .

Thus, if the region R is given by $r >> a$, it follows that *a static disclination in a body of infinite extent can exist in the absence of applied tractions at infinity.*

Clearly everything depends on knowing the region R in which the approximation (4.4.4) is valid. Now, $\underset{\sim}{Z}$ is known in R and (4.4.13) gives $||\underset{\sim}{Z}|| \approx g_o r^{-2}$. Since the stress is bounded in R, there exists a constant k in R such that $||\underset{\sim}{\chi}|| < k\, r$; simply note that $\chi^i = \delta^i_A X^A + u^i(X^A)$. Thus,

$$||\frac{1}{2s_2} \underset{\sim}{Z}\underset{\sim}{\gamma}_\alpha \underset{\sim}{\chi}|| \approx \frac{kg_o}{2s_2} r^{-1}$$

and hence the region R is given by $r >> \frac{kg_o}{2s_2}$. The approximate solution given by (4.4.5) and (4.4.13) is the far field solution of a static disclination in the absence of applied tractions at infinity. Thus, although the situation is significantly more complicated outside R (i.e., within a finite neighborhood of the singularity at $X^1 = X^2 = X^3 = 0$) useful additional information has been obtained about the nature of the far field of a disclination.

4.5 The Linear Elasticity Approximation Without Disclinations

We restrict our considerations in this section to material bodies for which disclinations are ignored from the start; that is, there is no breaking of the homogeneity of the action of the subgroup $SO(3)_0$. The relevant field equations are therefore those given in Section 3.15 for the Lagrangian $L = L_0 - s_1 L_1$. Next, we make the standard linearization whereby linear elasticity obtains from nonlinear elasticity:

$$\chi^i = \delta^i_A X^A + u^i(X^b) ,$$

$$E_{AB} \doteq \delta_{jA}(\partial_B u^j + \phi^j_B) + \delta_{jB}(\partial_A u^j + \phi^j_A) , \tag{4.5.1}$$

$$\sigma^A_i = \lambda\, \delta^A_i\, \delta^R_j(\partial_R u^j + \phi^j_R) + \mu(\delta^A_j\, \delta^R_i + \delta_{ij}\, \delta^{AR})(\partial_R u^j + \phi^j_R) .$$

It is then simply a matter of substituting (4.5.1) into the field equations (3.15.5,6) to obtain the governing equations in this approximation.

We have previously noted that any convenient system of gauge conditions may be used in the theory, although the antiexact gauge conditions lead to marked conceptual simplifications. In this inst nce, it turns out that the maximal simplification of the field equations is achieved through use of what may be termed the "pseudo-Lorentz" gauge conditions

$$(4.5.2) \qquad \delta^{AB}\,\partial_A\phi^i_B = \frac{1}{y}\,\partial_4\phi^i_4\ , \quad i = 1,2,3\ .$$

With the parameters

$$(4.5.3) \qquad c^2 = \frac{\mu}{\rho_o}\ , \quad \kappa^2 = \frac{\mu}{2s_1}\ , \quad L = \frac{\lambda}{\mu}\ , \quad a^2 = \frac{y}{c^2}\ ,$$

the field equations are given by

$$(4.5.4) \qquad (\nabla^2 - \frac{1}{y}\,\partial_4\partial_4 - \kappa^2)\phi^m_E - \kappa^2(\delta_{Ej}\ \delta^{Rm} + L\ \delta^m_E\ \delta^R_j)\phi^j_R$$
$$= \kappa^2(L\ \delta^m_E\ \partial_j + \delta_{Ej}\delta^{mk}\ \partial_k + \delta^m_j\ \delta^k_E\ \partial_k)u^j\ ,$$

$$(4.5.5) \qquad (\nabla^2 - \frac{1}{y}\,\partial_4\partial_4 - \kappa^2\ a^2)\phi^i_4 = \kappa^2\ a^2\ \partial_4 u^i\ ,$$

$$(4.5.6) \qquad (\nabla^2 - \frac{1}{c^2}\,\partial_4\partial_4)u^m + (L + 1)\delta^{mj}\ \partial_j\partial_i u^i$$
$$= [(a^2 - 1)\delta^m_j\ \delta^{RA} - L\ \delta^{mA}\ \delta^R_j - \delta^A_j\ \delta^{mR}]\partial_A\phi^j_R\ .$$

We first note that (4.5.4) and (4.5.6) do not involve the field variables ϕ^i_4 . Further, if we solve (4.5.4) and (4.5.6) for $(u^i,\ \phi^i_A)$, then (4.5.2) and (4.5.5) can be used to obtain ϕ^i_4 . We confine our attention from now on to the system consisting of (4.5.4) and (4.5.6).

The system (4.5.4) is a coupled system of Klein-Gordon equations. If we write $\underset{\sim}{\Phi} = ((\phi^m_E))$, and let $\underset{\sim}{\sigma}$

denote the matrix whose entries occur on the right-hand sides of (4.5.4), then (4.5.1) shows that $\underset{\sim}{\sigma}$ is $1/\mu$ times the matrix whose entries are the elastic Piola-Kirchhoff stress tensor that is calculated from the displacement gradients, $\partial_A u^i$, alone. With this notation, (4.5.4) becomes

$$(4.5.7) \qquad (\nabla^2 - \frac{1}{y}\partial_4\partial_4 - \kappa^2)\underset{\sim}{\Phi} - \kappa^2(\underset{\sim}{\Phi}^T + L\,\mathrm{Tr}(\underset{\sim}{\Phi})\underset{\sim}{I}) = \kappa^2\underset{\sim}{\sigma} .$$

Noting that $\underset{\sim}{\sigma}$ is symmetric, it follows immediately from (4.5.7) that $(\nabla^2 - \frac{1}{y}\partial_4\partial_4)(\underset{\sim}{\Phi} - \underset{\sim}{\Phi}^T) = \underset{\sim}{0}$. Having noted this possibility, we shall now put $\underset{\sim}{\Phi} = \underset{\sim}{\Phi}^T$ since antisymmetric $\underset{\sim}{\Phi}$'s make no contribution in any of the other field equations. $\underset{\sim}{\Phi}$ is thus given by

$$(4.5.8) \qquad \underset{\sim}{\Phi} = \underset{\sim}{\hat{\Phi}} + \frac{1}{3}\eta\,\underset{\sim}{I} ,$$

where $\underset{\sim}{\hat{\Phi}}$ is trace-free, and $\underset{\sim}{\hat{\Phi}}$ and η satisfy

$$(4.5.9) \qquad (\nabla^2 - \frac{1}{y}\partial_4\partial_4 - 2\kappa^2)\underset{\sim}{\hat{\Phi}} = \kappa^2(\underset{\sim}{\sigma} - \frac{1}{3}\mathrm{Tr}(\underset{\sim}{\sigma})\underset{\sim}{I}) ,$$

$$(4.5.10) \qquad (\nabla^2 - \frac{1}{y}\partial_4\partial_4 - 2\kappa^2(1 + \frac{3}{2}L))\eta = \kappa^2\,\mathrm{Tr}(\underset{\sim}{\sigma}) .$$

Thus, *the trace-free part of* $\underset{\sim}{\Phi}$ *is driven by the elastic stress deviatoric* $\underset{\sim}{\sigma} - \frac{1}{3}\mathrm{Tr}(\underset{\sim}{\sigma})\underset{\sim}{I}$.

The uncoupling of the ϕ-equations, as represented by (4.5.8)-(4.5.10), permits a direct solution of the

ϕ-equations since the scalar Klein-Gordon operator on an infinite spatial domain has a well defined Green's function for zero Cauchy data [35, p. 854] (a defect free reference state). The ϕ's can thus be written as a uniquely determined system of integro-differential operators acting on the variables $u^i(X^b)$. When these evaluations are substituted into the right-hand sides of (4.5.6), we obtain an explicit system of integro-differential equations for the u-fields. Thus, the presence of dislocations leads to a nonlocal system of field equations for the u-fields.

For a body of infinite spatial extent, the linearity of the field equations (4.5.4) and (4.5.6) permits direct use of Fourier transform methods. With the standard kernel $\exp(i(k_A X^A - \omega T))$, an elimination of the Fourier transforms of the ϕ's between (4.5.4) and (4.5.6) with the aid of (4.5.8)-(4.5.10) gives us the equations

(4.5.11) $$(P\,\delta^m_j + Q\,k_j k^m)\hat{U}^j = 0\ .$$

Here, $\hat{U}^j$ stands for the Fourier transform of u^j,
$k^2 = \delta^{AB} k_A k_B$,

(4.5.12) $$P = (\tfrac{\omega}{c})^2 - k^2 + \kappa^2(a^2 - 2)G_1^{-1} k^2\ ,$$

(4.5.13) $$Q = -1 - L - \kappa^2 L(2 + 3L)G_2^{-1}$$
$$+ \kappa^2(a^2 - 2)[(L + 1)G_1^{-1} + \tfrac{1}{3}(2 + 3L)(G_2^{-1} - G_1^{-1})]\ ,$$

(4.5.14) $$G_1 = (\frac{\omega}{c})^2 a^{-2} - k^2 - 2\kappa^2 ,$$

(4.5.15) $$G_2 = (\frac{\omega}{c})^2 a^{-2} - k^2 - \kappa^2(2 + 3L) .$$

Particular note should be taken of the strong decoupling that occurs for $a^2 = 2$ (i.e., when $y = 2c^2$).

It now follows directly from (4.5.11) that nontrivial solutions obtain only when

(4.5.16) $$0 = \det(P\,\delta_j^m + Q\,k_j k^m) = P^2(P + k^2 Q) .$$

Thus, as in classic linear elasticity, there is the double shear mode characterized by $P = 0$ and a simple longitudinal mode characterized by $P + k^2 Q = 0$. For the double shear mode, $P = 0$ gives the dispersion relation

$$((\frac{\omega}{c})^2 - k^2)((\frac{\omega}{c})^2 - a^2(k^2 + 2\kappa^2)) + \kappa^2 a^2 (a^2 - 2) k^2 = 0 ;$$

that is

(4.5.17) $$(\frac{\omega}{c})^2 = \frac{1}{2}(a^2 + 1)k^2 + a^2\kappa^2 \pm \frac{1}{2}\sqrt{(a^2 - 1)^2 k^4 + 4a^2\kappa^2 k^2 + 4a^4\kappa^4} .$$

When $y < c^2$ (i.e., for $a^2 < 1$) , we obtain dispersion curves that look like

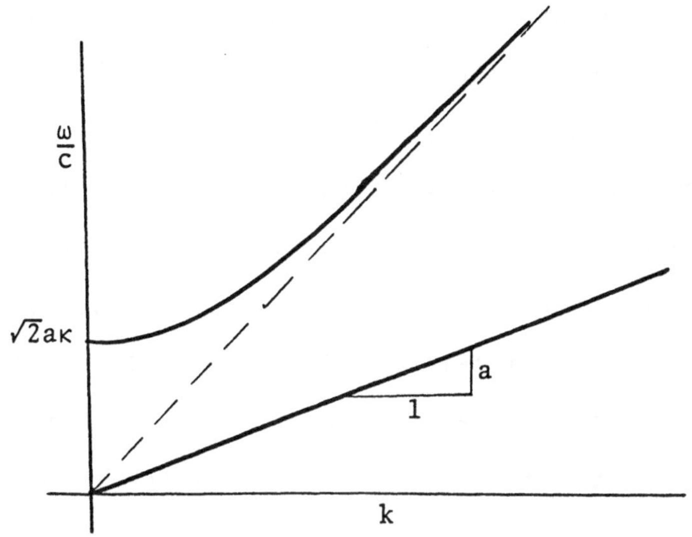

while for $y > c^2$ (i.e., for $a^2 > 1$), the dispersion curves look like

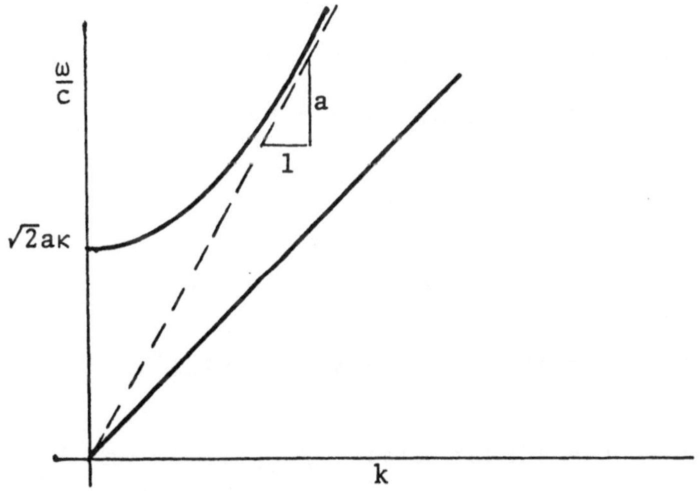

Since the speed of elastic shear waves should be approximately c with minor corrections for the interactions, it is clear that y must be greater than c. However, since y only determines the very short wavelength speed, this does not mean that dislocations propogate faster than shear waves. An inspection of the figure for $\sqrt{y} > c$ shows that the group velocity of dislocations for small k (for long wavelengths) is significantly less than the group velocity for elastic shear waves, as is actually known to be the case from experiments. It is also significant to note that the vertical intercept of the optical branch and the curvature of the optical branch at $k = 0$ serve to determine the two constants s_1 and y of the theory.

The gauge conditions used in the analysis of this problem were the pseudo-Lorentz ones, (4.5.2), rather than the antiexact gauge conditions. Accordingly, the field variables u^i may not be identified with the total integrable displacements of the body. This is easily seen by noting that the distortion 1-forms are given by

$$B^i = d(\delta^i_A X^A + u^i) + \phi^i ,$$

and $\phi^i = dH(\phi^i) + H(d\phi^i)$ can be used to obtain

$$B^i = d(\delta^i_A X^A + u^i + H(\phi^i)) + H(d\phi^i) ,$$

where H denotes the linear homotopy operator on E_4. The uniqueness of the decomposition $\phi^i = dH(\phi^i) + H(d\phi^i)$ then shows that the total integrable displacement field, u^i_{tot}, is given by

$$u^i_{tot} = u^i + H(\phi^i) \ .$$

In like manner, since the Burgers vector of a circuit, $b^i(\partial S_2)$, is given by

$$b^i(\partial S_2) = \int_{\partial S_2} \phi^i = \int_{\partial S_2} \{dH(\phi^i) + H(d\phi^i)\}$$

and the circuit integral of any regular total differential vanishes identically, we obtain

$$b^i(\partial S_2) = \int_{\partial S_2} H(d\phi^i) \ .$$

The resulting expression

$$B^i = d(\delta^i_A \, X^A + u^i_{tot}) + H(d\phi^i)$$

is thus uniquely decomposed in terms of the total integrable response $\chi^i_{tot} = \delta^i_A \, X^A + u^i_{tot}$ and the part $H(d\phi^i)$ that accounts for the dislocations. When defects are absent or when we use the antiexact gauge, the field variables u^i and u^i_{tot} coincide, otherwise they are distinct sets of quantities. Although this point has been made before, it requires repeated underscoring since *the actual displacements of a material body in the laboratory are quantified by the variables* u^i_{tot} .

4.6 Static Problems - Edge and Screw Dislocations

The problems just dealt with were concerned primarily with the linearized interaction between the elastic fields and the dislocation fields in the dynamic case. We can, however, consider what may be termed linearized dislocation problems by simply putting all of the elastic displacement variables, $u^i(X^a)$, to zero. As with most theories, the characteristic phenomena are described by constructing solutions with singularities. We accordingly relax the requirement that the reference state be defect free. If, in addition, we restrict our attention to the static case, the field variables $\phi^i_a(X^A, T)$ that describe the dislocations become independent of the time variable, T , and $\phi^i_4(X^A) = 0$. With the parameters

$$(4.6.1) \qquad \kappa^2 = \mu/2s_1 \ , \quad L = \lambda/\mu$$

and the notation $\partial^A = \delta^{AB}\partial_B$, the field equations (4.5.4)-(4.5.6) reduce to

$$(4.6.2) \qquad \nabla^2\phi^m_E - \partial_E(\partial^A\phi^m_A) = \kappa^2 \Sigma^m_E \ , \quad \partial^E \Sigma^m_E = 0 \ ,$$

$$(4.6.3) \qquad \Sigma^m_E = L(\delta^A_i \phi^i_A)\delta^m_E + \phi^m_E + \delta_{Ei} \delta^{mA} \phi^i_A \ ,$$

where the assocaited stress tensor is given by

$$(4.6.4) \qquad \sigma^m_E = \mu\Sigma^m_E \ .$$

Here, we have explicitly relaxed the pseudo-Lorentz gauge conditions $\partial^A \phi^i_A = 0$ used in the previous section. This accounts for the presence of the terms $\partial_E(\partial^A \phi^i_A)$ on the left-hand side of (4.6.2) .

Let $\vec{e}_1$, $\vec{e}_2$, $\vec{e}_3$ be the standard unit basis vectors associated with the Cartesian coordinate cover of the reference configuration, and let the vector fields $\vec{\phi}^i$ and $\vec{\Sigma}^i$ be defined by

$$(4.6.5) \qquad \vec{\phi}^i = \phi^i_A \, \vec{e}_A \ , \qquad \vec{\Sigma}^i = \Sigma^i_A \, \vec{e}_A \ .$$

It is then a simple matter to see that the field equations (4.6.2) are equivalent to the vector equations

$$(4.6.6) \qquad \vec{\nabla} \times \vec{\nabla} \times \vec{\phi}^i = - \kappa^2 \, \vec{\Sigma}^i \ , \quad \vec{\nabla} \cdot \vec{\Sigma}^i = 0 \ .$$

Standard results from magnetostatics then show that (4.6.6) are solvable for any vectors $\vec{\Sigma}^i$ that satisfy the continuity equations $\vec{\nabla} \cdot \vec{\Sigma}^i = 0$.

A plan of solution is thus clear. First, find classes of solutions of the continuity equations $\vec{\nabla} \cdot \vec{\Sigma}^i = 0$. Next, solve (4.6.3) for the ϕ's in terms of the Σ's . Finally, obtain the ϕ's as explicit functions of position by solving the integrable system of field equations given by the first of (4.6.6).

Up to this point, the gauge has been arbitrary. We may thus impose the explicit gauge conditions

(4.6.7) $\phi_2^1 = \phi_1^2 , \quad \phi_3^1 = \phi_1^3 , \quad \phi_3^2 = \phi_2^3 .$

This has the definite advantage of reducing (4.6.3) to

(4.6.8) $\Sigma_A^m = L(\delta_i^B \phi_B^i)\delta_A^m + 2\,\phi_A^m ,$

and hence

$$2\,\phi_A^m = \Sigma_A^m - \frac{L}{3L+2}(\delta_i^B \Sigma_B^i)\delta_A^m .$$

Thus, step two of the solution process becomes trivial. For simplicity in later calculations, we introduce the parameter

(4.6.9) $a = \frac{L}{3L+2} ,$

in which case we have

(4.6.10) $2\,\phi_A^m = \Sigma_A^m - a(\delta_i^B \Sigma_B^i)\delta_A^m .$

We confine our attention in the remainder of this Section to problems with axial orientation. The axis is taken to be the Z-axis and we set $r^2 = X^2 + Y^2$. A convenient notation here is achieved by allowing $\underset{\sim}{\Sigma}$ and $\underset{\sim}{\Phi}$ to denote the symmetric matrices whose entries are Σ_A^m and ϕ_A^m , respectively. It is then clear that

$$(4.6.11) \qquad \underset{\sim}{\Sigma} = \begin{pmatrix} 0 & 0 & -\partial_Y \\ 0 & 0 & \partial_X \\ -\partial_Y & \partial_X & 0 \end{pmatrix} f(r)$$

satisfies the equilibrium equations (continuity conditions), $\partial^A \Sigma^i_A = 0$ (i.e., $\vec{\nabla}\cdot\vec{\Sigma}^i = 0$). Since $\underset{\sim}{\Sigma}$ is trace free in this case, (4.6.10) gives $2\underset{\sim}{\phi} = \underset{\sim}{\Sigma}$, and hence it remains only to satisfy the field equations $\vec{\nabla} \times \vec{\nabla} \times \vec{\phi}^i = - 2\kappa^2 \vec{\phi}^i$. A simple calculation shows that the field equations are satisfied whenever $f(r)$ satisfies $\nabla^2 f = 2\kappa^2 f$; that is

$$(4.6.12) \qquad r \frac{d^2 f}{dr^2} + \frac{df}{dr} - 2\kappa^2 \, r \, f = 0 \, .$$

Let us confine our attention to solutions of (4.6.12) that remain bounded for large values of r . The most general solution of (4.6.12) with this property is $C_1 K_o(\sqrt{2}\kappa r)$, where K_o is the modified Bessel function of the second kind of order zero. When this is put back into (4.6.11), we have

$$(4.6.13) \qquad \underset{\sim}{\Sigma} = 2\underset{\sim}{\Phi} = \sqrt{2} C_1 \kappa r^{-1} \, K_1(\sqrt{2}\kappa r) \begin{pmatrix} 0 & 0 & -Y \\ 0 & 0 & X \\ -Y & X & 0 \end{pmatrix}$$

In the near field, $\sqrt{2}\kappa r \ll 1$, $K_1(\sqrt{2}\kappa r) \approx (\sqrt{2}\kappa r)^{-1}$, and hence (4.6.13) gives

$$(4.6.14) \qquad \underset{\sim}{\Sigma} = 2\underset{\sim}{\Phi} \approx C_1 r^{-2} \begin{pmatrix} 0 & 0 & -Y \\ 0 & 0 & X \\ -Y & X & 0 \end{pmatrix} , \quad \sqrt{2}\kappa r \ll 1 \, .$$

This agrees exactly with the stress distribution for a screw dislocation along the Z-axis [20, p. 57] if we take the constant C_1 to be given by

(4.6.15) $$C_1 = b/2\pi$$

(recall that $\underset{\sim}{\sigma} = \mu\underset{\sim}{\Sigma}$) . On the other hand, the far field expansion ($\sqrt{2}\kappa r>>1$) gives $K_1(\sqrt{2}\kappa r) \approx \sqrt{\frac{\pi}{2\sqrt{2}\kappa r}}\, \exp(-\sqrt{2}\kappa r)$ and hence, with (4.6.15), we have

(4.6.16) $$\underset{\sim}{\Sigma} = 2\underset{\sim}{\Phi} \approx \frac{b\sqrt{\kappa}}{2\sqrt{\sqrt{2}\pi}}\, r^{-3/2} \exp(-\sqrt{2}\kappa r) \begin{pmatrix} 0 & 0 & -Y \\ 0 & 0 & X \\ -Y & X & 0 \end{pmatrix} .$$

It sould be noted that the near field solution is independent of the coupling constant κ while the far field solution depends explicitly on κ and decays exponentially rather than simply as r^{-1} for the far field of the classic screw dislocation. If this solution is to be applied in the context of an underlying periodic crystaline structure, the periodicity of the crystal simply continues the near field solution through the body. From another point of view, the classic theory obtains in the limit when s_1 becomes very large; that is, when κ tends to zero. Accordingly, since the near field representation is valid for $\sqrt{2}\kappa r<<1$, we obtain agreement with the classic solution in the limit as κ tends to zero. Of course, the solution has no physical meaning in a sufficiently small neighborhood of $r = 0$ since the assumptions underlying the linear engineering approximation are violated.

Perhaps a better understanding of the solution can be obtained by computing Burgers vectors directly in terms of the ϕ's . For this purpose, we consider a closed circular arc in any plane Z=constant with center on the Z axis and of radius R . Since $\phi^i = \phi^i_A \, dX^A$, (4.6.13) and (4.6.15) show that $\phi^1 = \phi^2 = 0$ and

(4.6.17) $$\phi^3 = \frac{b}{4\pi} \sqrt{2}\kappa R^{-1} \, K_1(\sqrt{2}\kappa R)(-Y \, dX + X \, dY)$$

for this circular arc on the plane Z = constant. Thus, since the Burgers vector for a circuit is given by $b^i = \oint \phi^i$, we obtain $b^1 = b^2 = 0$ and

(4.6.18) $$b^3 = \frac{b}{2} \sqrt{2}\kappa R \, K_1(\sqrt{2}\kappa R) \; .$$

Thus, in the near field we obtain

(4.6.19) $$b^3 \approx \frac{b}{2} \; , \quad \sqrt{2}\kappa R \ll 1 \; ,$$

while in the far field

(4.6.20) $$b^3 = \frac{b}{2\sqrt{2}} \sqrt{\sqrt{2}\kappa R} \, \exp(-\sqrt{2}\kappa R) \; , \quad \sqrt{2}\kappa R \gg 1 \; .$$

The full solution given by (4.6.14) and (4.6.15) thus describes a classic screw dislocation in the neighborhood of the Z axis that becomes lost to external observers at sufficiently large distances from the Z axis.

The next class of problems is characterized by

$$(4.6.21) \qquad \underset{\sim}{\Sigma} = \begin{pmatrix} \partial_Y\partial_Y f & -\partial_Y\partial_X f & 0 \\ -\partial_Y\partial_X f & \partial_X\partial_X f & 0 \\ 0 & 0 & p \end{pmatrix}$$

where the functions f and p depend only on the variables X and Y only. It is clear that (4.6.21) satisfies the symmetry conditions $\underset{\sim}{\Sigma} = \underset{\sim}{\Sigma}^T$ and the equilibrium conditions $\vec{\nabla}\cdot\vec{\Sigma}^i = 0$. It then follows directly from (4.6.10) that

$$(4.6.22) \qquad 2\underset{\sim}{\Phi} = \begin{pmatrix} \partial_Y\partial_Y f - a\nabla^2 f - ap & -\partial_Y\partial_X f & 0 \\ -\partial_Y\partial_X f & \partial_X\partial_X f - a\nabla^2 f - ap & 0 \\ 0 & 0 & (1-a)p - a\nabla^2 f \end{pmatrix} .$$

A direct calculation then shows that

$$(4.6.23) \qquad \partial^A\phi^1_A = -a\partial_X(\nabla^2 f + p) , \quad \partial^A\phi^2_A = -a\partial_Y(\nabla^2 f + p) ,$$

$$\partial^A\phi^3_A = 0$$

When (4.6.21)-(4.6.23) are substituted into the field equations (4.6.2), we obtain the conditions

$$(4.6.24) \qquad (1 - a)\nabla^2 f - 2\kappa^2 f = a\,p ,$$

$$(4.6.25) \qquad (1 - a)\nabla^2 p - 2\kappa^2 p = a\nabla^4 f .$$

The first of these serves to determine p in terms of f through

$$(4.6.26) \qquad p = \frac{1}{a}\{(1 - a)\nabla^2 f - 2\kappa^2 f\} .$$

When this is used to eliminate p from (4.6.25), we obtain the following equation for the determination of f :

$$(4.6.27) \qquad (\nabla^2 - M^2)(\nabla^2 - N^2)f = 0 .$$

Here, the parameters M and N are related to the previous parameters by

$$(4.6.28) \qquad M^2 = 2\kappa^2 , \quad N^2 = 2\kappa^2 \frac{1}{1-2a} = 2\kappa^2 \frac{3L+2}{L+2} .$$

The solutions of (4.6.27) are clearly more complex than those of the previous problem in which we only had to solve $(\nabla^2 - M^2)f = 0$. However, in the near field, that is, for $\kappa r<<1$, it is sufficient to solve the equation that obtains from allowing κ to tend to zero (i.e., M and N tending to zero);

$$(4.6.29) \qquad \nabla^4 f = 0 , \quad \kappa r<<1 .$$

Solutions of (4.6.29) are given by

$$(4.6.30) \qquad f = X\, u(X,Y) + Y\, v(X,Y) + w(X,Y)$$

where $w(X,Y)$ is a harmonic function and $u(X,Y)$ and

$v(X,Y)$ are conjugate harmonic functions. It then follows directly from (4.6.21), (4.6.29) and (4.6.30) that the stress distribution in near field agrees with that predicted by the known solutions for edge dislocations [20,p. 56], simply take $f = C_1 Y \ln(r)$ or $f = C_2 X \ln(r)$. The near field solutions of our field equations thus reproduce the classic edge dislocation solutions. The same remarks as given previously apply here as well. In particular, in a crystaline substance, the crystal periodicity will have the effect of periodically continuing the near field solution so that it applies throughout the crystaline body.

For the far field, $r \gg 1$, (4.6.27) admits the approximation

$$(4.6.31) \qquad \left(\frac{d^2}{dr^2} - M^2\right)\left(\frac{d^2}{dr^2} - N^2\right) f(r) = 0 .$$

There is thus a 2-parameter family of solutions whose behavior at infinity is dominated by exponential decay. In fact, it should now be clear that one of the basic effects of the gauge theory of static dislocations arises primarily through the replacement of the operator ∇^2 by the operator $\nabla^2 - M^2$.

It is to be specifically noted that the theory has reproduced the stress fields associated with edge and screw dislocations without any consideration of accompanying displacement fields. In fact, the field variables u^i have been put identically equal to zero. There is thus a useful parallel that can be drawn. The classic theory of dislocations describes dislocations by specifying either a multiple valued or discontinuous displacement boundary value problem. The linear gauge theory of dislocations formulates the problem as a traction boundary value problem and obtains the stress distributions directly. From this viewpoint, the field equations $\nabla^2\phi^m_E - \partial_E(\partial^A\phi^m_A) = \kappa^2\, \Sigma^m_E$ of the gauge theory are what replace the compatibility conditions of linear elasticity theory.

Apart from replicating the stress distributions associated with edge dislocations, the results obtained above also prove to be useful for a wide range of problems. We first note that although we have eliminated the variables u^i, we have not imposed the antiexact gauge condition. Thus, for any given solution ϕ^i_A of the field equations, the totally integrable displacement fields are recovered by the assignment

$$\text{(4.6.32)} \qquad \chi^i_{total} = u^i(X^B) + \delta^i_A\, X^A = H(\phi^i_E\, dX^E)$$

since $B^i_A\, dX^A = \phi^i_A\, dX^A = dH(\phi^i_E\, dX^E) + H(d\phi^i_E \wedge dX^E)$. Further, if we allow the function f (and hence the function p) to be a function of X and Y , rather than $r^2 = X^2 + Y^2$, then (4.6.21) reproduces the stress field that would be naturally associated with problems of plane strain. There is a difference, however, for (4.6.24) and (4.6.21) show that the σ_3^3 component differs from that of classic elasticity theory for $\kappa^2 \neq 0$. In addition, the function f must now satisfy (4.6.31) instead of the biharmonic equation of classic elasticity theory. These differences notwithstanding, the equations derived in this section provide direct access to the class of plane strain problems in the presence of dislocations.

4.7 Traction Boundary Value Problems - Expansion in the Load Parameter

Let B be a finite body whose boundary, ∂B , is loaded by a given system of static surface tractions $\{T_i\}$, and let $\{S_i^A , \hat{u}^i\}$ be the components of the Piola-Kirchhoff stress tensor and the displacement functions that solve the corresponding traction boundary value problem of *linear* elasticity. The applied tractions are assumed to be piecewise smooth and bounded on ∂B so that there exists a numerical parameter T with the dimensions of force per unit area such that $T_i = O(T)$. Let us confine our attention to problems where ∂B is sufficiently smooth that there are no unbounded stress concentrations. Under these conditions we can write

$$(4.7.1)\qquad S_i^A = \mu\hat{S}_i^A , \quad T_i = \mu\hat{T}_i ,$$

$$\hat{S}_i^A = O(\varepsilon) , \quad \hat{T}_i = O(\varepsilon) , \quad \partial_A\hat{u}^i = O(\varepsilon) ,$$

where

$$(4.7.2)\qquad \varepsilon = \frac{T}{\mu}$$

is the *load parameter* associated with the linear problem.

Because the body is finite and subject to specified surface tractions, dislocated but disclination free states of the body are described by the field equations derived in Section 3.18. We confine our attention throughout this Section to static problems, so that $\phi^i_4 = 0$ and $\{\chi^i, \phi^i_A\}$ depend only on the reference configuration coordinates (X^A). The governing field equations are therefore given by

$$(4.7.3) \qquad \partial_A \hat{\sigma}^A_i = 0 \ , \quad \partial_A \hat{S}^A_i = 0 \ ,$$

$$(4.7.4) \qquad \delta_{ji} \ \delta^{BD} \ \partial^A D^i_{AD} = \frac{\mu}{2s_1} \ (\hat{\sigma}^B_j - \hat{S}^B_j) \ ,$$

where s_1 is the dislocation coupling constant, and the boundary conditions are

$$(4.7.5) \qquad \hat{\sigma}^A_j \Big|_{\partial B} N_A = \hat{T}_j = \hat{S}^A_j \Big|_{\partial B} N_A \ , \quad D^i_{AB} \Big|_{\partial B} N^A = 0 \ .$$

Here, the constitutive relations are $(4.7.1)_1$ and

$$(4.7.6) \qquad \sigma^A_i = \mu \ \hat{\sigma}^A_i \ , \ D^i_{AB} = \partial_A \phi^i_B - \partial_B \phi^i_A \ ,$$

$$(4.7.7) \qquad B^i_A = \partial_A \chi^i + \phi^i_A = \delta^i_A + \partial_A u^i + \phi^i_A \ ,$$

$$(4.7.8) \qquad \hat{\sigma}^A_i = \frac{1}{2} \ \delta^A_B \ \delta_{ij} \ B^j_C \ (\frac{\lambda}{\mu} \delta^{BC} \delta^{FD} E_{FD} + 2\delta^{RB} \delta^{SC} E_{RS}) \ ,$$

$$(4.7.9) \qquad E_{AB} = B^i_A \ \delta_{ij} \ B^j_B - \delta_{AB} \ ,$$

$$(4.7.10) \qquad \hat{S}^A_i = \frac{1}{2} \ \delta^A_B \ \delta_{iC} (\frac{\lambda}{\mu} \delta^{BC} \delta^{FD} \hat{E}_{FD} + 2\delta^{RB} \delta^{SC} \hat{E}_{RS}) \ ,$$

(4.7.11) $\hat{E}_{AB} = \delta_{Ai}\,\partial_B \hat{u}^i + \delta_{Bi}\,\partial_A \hat{u}^i$.

An examination of these equations shows that we can effect an expansion in the load parameter ε provided an explicit statement is made concerning the order of μ/s_1 . In contrast with Section 4.2, we consider problems for which

(4.7.12) $\mu/s_1 = O(1)$.

Thus, we study problems whose solutions may be expressed by

$$\hat{\sigma}^A_i = \varepsilon \underset{1}{\sigma}{}^A_i + \varepsilon^2 \underset{2}{\sigma}{}^A_i + \varepsilon^3 \underset{3}{\sigma}{}^A_i + \ldots ,$$

$$\hat{S}^A_i = \varepsilon \underset{1}{S}{}^A_i , \quad \hat{T}_i = \varepsilon \underset{1}{T}{}_i , \quad \partial_A \hat{u} = \varepsilon\, \partial_A \bar{u}^i ,$$

(4.7.13) $$B^i_A = \varepsilon \underset{1}{B}{}^i_A + \varepsilon^2 \underset{2}{B}{}^i_A + \varepsilon^3 \underset{3}{B}{}^i_A + \ldots ,$$

$$D^i_{AB} = \varepsilon \underset{1}{D}{}^i_{AB} + \varepsilon^2 \underset{2}{D}{}^i_{AB} + \varepsilon^3 \underset{3}{D}{}^i_{AB} + \ldots ,$$

$$E_{AB} = \varepsilon \underset{1}{E}{}_{AB} + \varepsilon^2 \underset{2}{E}{}_{AB} + \varepsilon^3 \underset{3}{E}{}_{AB} + \ldots ,$$

where all quantities with indices below them are $O(1)$ throughout B . It is then a simple matter to see that the governing equations break up into the following collections:

$$\partial_A \underset{1}{S}{}^A_i = 0 \ , \quad \underset{1}{S}{}^A_i \Big|_{\partial B} N_A = \underset{1}{T}_i \ ;$$

(4.7.14)
$$\partial_A \underset{1}{\sigma}{}^A_i = 0 \ , \quad \underset{1}{\sigma}{}^A_i \Big|_{\partial B} N_A = \underset{1}{T}_i \ ;$$

$$\delta_{ji} \ \delta^{BD} \ \partial^A \underset{1}{D}{}^i_{AD} = \frac{\mu}{2s_1} (\underset{1}{\sigma}{}^B_j - \underset{1}{S}{}^B_j) \ , \quad \underset{1}{D}{}^i_{AB} \Big|_{\partial B} N^A = 0$$

from the first order equations and

$$\partial_A \underset{r}{\sigma}{}^A_i = 0 \ , \quad \underset{r}{\sigma}{}^A_i \Big|_{\partial B} N_A = 0 \ ;$$

(4.7.15)
$$\delta_{ji} \ \delta^{BD} \partial^A \underset{r}{D}{}^i_{AD} = \frac{\mu}{2s_1} \underset{r}{\sigma}{}^B_j \ , \quad \underset{r}{D}{}^i_{AB} \Big|_{\partial B} N^A = 0$$

for the higher order equations.

A direct combination of the equations listed above shows that the sytem (4.7.14) is equivalent to

(4.7.16)
$$\partial_A (\underset{e}{Y}{}^A_i + \underset{\phi}{Y}{}^A_i) = 0 \ , \quad (\underset{e}{Y}{}^A_i + \underset{\phi}{Y}{}^A_i) \Big|_{\partial B} N_A = 0 \ ,$$

(4.7.17)
$$\delta_{ji} \ \delta^{BD} \ \partial^A \underset{1}{D}{}^i_{AD} = \frac{\mu}{2s_1} (\underset{e}{Y}{}^B_j + \underset{\phi}{Y}{}^B_j) \ , \quad \underset{1}{D}{}^i_{AB} \Big|_{\partial B} N^A = 0 \ ,$$

where

$$\underset{e}{Y}{}^A_i = \frac{1}{2} \delta^A_B \ \delta_{iC} (\frac{\lambda}{\mu} \delta^{BC} \delta^{FD} \underset{e}{E}_{FD} + 2\delta^{RB} \delta^{SC} \underset{e}{E}_{RS}) \ ,$$

(4.7.18)
$$\underset{\phi}{Y}{}^A_i = \frac{1}{2} \delta^A_B \ \delta_{iC} (\frac{\lambda}{\mu} \delta^{BC} \delta^{FD} \underset{\phi}{E}_{FD} + 2\delta^{RB} \delta^{SC} \underset{\phi}{E}_{RS}) \ ,$$

$$(4.7.19)\quad \begin{aligned} \underset{e}{E}_{AB} &= \delta_{Ai}\,\partial_B(\underset{1}{u}^i - \bar{u}^i) + \delta_{Bi}\,\partial_A(\underset{1}{u}^i - \bar{u}^i)\ , \\ \underset{\phi}{E}_{AB} &= \delta_{Ai}\,\underset{1}{\phi}^i_B + \delta_{Bi}\,\underset{1}{\phi}^i_A\ ,\quad \underset{1}{D}^i_{AB} = \partial_A\,\underset{1}{\phi}^i_B - \partial_B\,\underset{1}{\phi}^i_A \end{aligned}$$

A solution of this system subject to the gauge conditions $X^A\underset{1}{\phi}^i_A = 0$ is given by

$$(4.7.20)\quad \underset{1}{u}^i = \bar{u}^i + k^i\ ,\quad \underset{1}{\phi}^i_A = 0\ ,\quad \underset{1}{\sigma}^A_i = \underset{1}{S}^A_i\ ,$$

and it is the only solution under the moderate smoothness conditions that are sufficient for uniqueness of the homogeneous boundary value problem (4.7.16)-(4.7.19) and $X^A\phi^i_A = 0$. Thus, the first order response is that predicted by linear elasticity theory and dislocation effects do not obtain until at least the second order. This result is similar to that obtained in Section 4.2 although the ε-order condition on the dislocation coupling constant is weaker ($\mu/s_1 = O(1)$ rather than $\mu/s_1 = O(\varepsilon)$) .

The result just obtained for the first order solution leads to significant simplifications of the equations in succeeding orders. We first note that (4.7.7) through (4.7.9) and (4.7.13) now yield the evaluations

$$(4.7.21)\quad \underset{2}{\sigma}^A_i = \frac{1}{2}\,\delta_{ij}\{\frac{\lambda}{\mu}\delta^{Aj}\delta^{FD}\,\underset{2}{E}_{FD} + 2\delta^{RA}\delta^{Sj}\,\underset{2}{E}_{RS} + \partial_C\bar{u}^j(\frac{\lambda}{2}\delta^{AC}\delta^{FD}\,\bar{E}_{FD} + 2\delta^{RA}\delta^{SC}\bar{E}_{RS})\}\ ,$$

$$(4.7.22)\qquad \underset{2}{E}_{AB} = \delta_{Aj}(\partial_B \underset{2}{u}^j + \underset{2}{\phi}^j_B) + \delta_{Bj}(\partial_A \underset{2}{u}^j + \underset{2}{\phi}^j_A) + \partial_A\bar{u}^i\,\delta_{ij}\,\partial_B\bar{u}^j$$

$$(4.7.23)\qquad \underset{3}{\sigma}^A_i = \frac{1}{2}\,\delta_{ij}\{\frac{\lambda}{\mu}\delta^{Aj}\delta^{FD}\,\underset{3}{E}_{FD} + 2\delta^{RA}\delta^{Sj}\,\underset{3}{E}_{RS} + \partial_C\bar{u}^j(\frac{\lambda}{\mu}\delta^{AC}\delta^{FD}\,\underset{2}{E}_{FD} + 2\delta^{RA}\delta^{SC}\,\underset{2}{E}_{RS}) + (\partial_C\underset{2}{u}^j + \underset{2}{\phi}^j_C)(\frac{\lambda}{\mu}\delta^{AC}\delta^{FD}\,\bar{E}_{FD} + 2\delta^{RA}\delta^{SC}\,\bar{E}_{RS})\}$$

$$(4.7.24)\qquad \underset{3}{E}_{AB} = \delta_{Aj}(\partial_B \underset{3}{u}^j + \underset{3}{\phi}^j_B) + \delta_{Bj}(\partial_A \underset{3}{u}^j + \underset{3}{\phi}^j_A)\quad \partial_A\bar{u}^i\,\delta_{ij}(\partial_B\underset{2}{u}^j + \underset{2}{\phi}^j_B) + (\partial_A\underset{2}{u}^i + \underset{2}{\phi}^i_A)\delta_{ij}\,\partial_B\bar{u}^j$$

with similar expressions for the higher order stresses and strain measures.

The field equations (4.7.15) provide the basis for establishing a result that was previously known only for dislocated bodies with traction free boundaries: *the mean value of each component of the* r-th *order stress vanishes for all* $r>1$. We simply integrate $(4.7.15)_1$ over the body B and then use the divergence theorem and the boundary conditions $(4.7.15)_2$ to obtain

$$(4.7.25)\qquad \int_B \underset{r}{\sigma}^B_j\,dV = 0\ .$$

In fact, the result is true in general if we replace the r-th order stress by the total effective stress. In order to see this, we integrate the field equations (4.7.4) over the whole body and then use the divergence theorem and the boundary conditions $(4.7.5)_2$ to obtain

$$(4.7.26) \qquad \int_B (\sigma_j^B - S_j^B)\,dV = 0 \; .$$

An inspection of (4.7.21)-(4.7.24) shows that

$$(4.7.27) \qquad \underset{r}{\sigma}{}_j^B = \underset{r}{Z}{}_j^B + C_{jk}^{BF}(\partial_F \underset{r}{u}{}^k + \underset{r}{\phi}{}_F^k)$$

where

$$C_{jk}^{BF} = \frac{\lambda}{\mu}\, \delta_j^B\, \delta_k^F + \delta_k^B\, \delta_j^F + \delta_{jk}\, \delta^{BF}$$

are the components of the complience tensor of linear elasticity and the quantities denoted by the Z's are polynomial functions of the solutions of lower order equations and are thus known functions of the reference configuration variables (X^A). The σ's of r-th order are thus linear functions of the unknowns $(\underset{r}{u}{}^i, \underset{r}{\phi}{}_A^i)$ and hence the field equations (4.7.15) of r-th order are linear inhomogeneous partial differential equations with driving terms $\underset{r}{Z}{}_j^B$. If we make the substitutions

$$(4.7.28) \qquad \underset{r}{u}{}^i = \underset{r}{v}{}^i + \underset{r}{\psi}{}^i \, , \quad \underset{r}{\phi}{}_A^i = \underset{r}{\psi}{}_A^i - \partial_A \underset{r}{\psi}{}^i$$

then

$$\partial_F \underset{r}{u}^k + \underset{r}{\phi}^k_F = \partial_F \underset{r}{v}^k + \underset{r}{\psi}^k_F$$

and

$$\underset{r}{D}^i_{AB} = \partial_A \underset{r}{\phi}^i_B - \partial_B \underset{r}{\phi}^i_A = \partial_A \underset{r}{\psi}^i_B - \partial_B \underset{r}{\psi}^i_A ,$$

so that the r-th order field equations (4.7.15) are form invariant under the gauge transformations (4.7.28) for any choice of the generating functions $\underset{r}{\psi}^i$. This fact can be used to secure satisfaction of the gauge conditions

(4.7.29) $$\partial^A \underset{r}{\psi}^i_A = 0$$

through the choice

(4.7.30) $$\nabla^2 \underset{r}{\psi}^i = \partial^A \underset{r}{\phi}^i_A ,$$

in which case we have the marked simplification

(4.7.31) $$\partial^A \underset{r}{D}^i_{AD} = \nabla^2 \underset{r}{\psi}^i_D .$$

Let H be the linear homotopy operator with center at a point P within the body B . If the original ϕ's satisfy the antiexact gauge condition

$$H(\underset{r}{\phi}^i_A \, dX^A) = 0 ,$$

so that the u's are the *total* displacement fields, then the second of (4.7.28) gives

$$\underset{r}{\psi}^i = H(\underset{r}{\psi}^i_A \, dX^A) \ .$$

The first of (4.7.28) then gives the following evaluation of the total displacement field

$$\underset{r}{u}^i = \underset{r}{v}^i + H(\underset{r}{\psi}^i_A \, dX^A) \ . \tag{4.7.32}$$

The r-th order approximation is thus governed by the linear inhomogeneous field equations

$$\partial^A \underset{r}{\psi}^i_A = 0 \ , \tag{4.7.33}$$

$$- C^{BF}_{jk} \, \partial_B (\partial_F \, \underset{r}{v}^k + \underset{r}{\psi}^k_F) = \partial_B \, \underset{r}{Z}^B_j \ , \tag{4.7.34}$$

$$\delta_{ji} \, \delta^{BD} \, \nabla^2 \underset{r}{\psi}^i_D - \frac{\mu}{2s_1} \, C^{BF}_{jk} (\partial_F \, \underset{r}{v}^k + \underset{r}{\psi}^k_F) = \frac{\mu}{2s_1} \, \underset{r}{Z}^B_j \tag{4.7.35}$$

and the boundary conditions

$$N_B \, C^{BF}_{jk} (\partial_F \, \underset{r}{v}^k + \underset{r}{\psi}^k_F) \Big|_{\partial B} = - N_B \, \underset{r}{Z}^B_j \Big|_{\partial B} \ , \tag{4.7.36}$$

$$\overset{\bullet}{N}{}^A (\partial_A \, \underset{r}{\psi}^i_B - \partial_B \, \underset{r}{\psi}^i_A) \Big|_{\partial B} = 0 \ . \tag{4.7.37}$$

An inspection of the above system shows that any functions $(\underset{r}{v}^i, \underset{r}{\psi}^i_A)$ that satisfy the field equations

(4.7.33), (4.7.35) and the boundary conditions (4.7.36), (4.7.37) will also satisfy the field equations (4.7.34) identically. We thus have a primitively deterministic system of 3 + 9 field equations (4.7.33), (4.7.35) for which the 3 + 6 boundary conditions (4.7.36), (4.7.37) serve to determine the solutions to within homogeneous rigid body translations $\underset{r}{v}^i \mapsto \underset{r}{v}^i + \underset{r}{k}^i$. Simply note that the field equations and the boundary conditions are invariant under homogeneous rigid body translation (the derivatives of the v's are involved but not the v's themselves).

An alternative and often useful choice of gauge is that obtained by choosing

(4.7.38) $$\underset{r}{\psi}^i = \underset{r}{u}^i ,$$

in which case (4.7.28) yields

(4.7.39) $$\underset{r}{v}^i = 0 , \qquad \underset{r}{\psi}^i_A = \underset{r}{\phi}^i_A + \partial_A \underset{r}{u}^i .$$

The displacement fields are then recovered by

(4.7.40) $$\underset{r}{u}^i = H(\underset{r}{\psi}^i_A \, dX^A) .$$

In this event, the r-th order field equations are

(4.7.41) $$- C^{BF}_{jk} \, \partial_B \underset{r}{\psi}^k_F = \partial_B \underset{r}{Z}^B_J ,$$

(4.7.42) $$\delta_{ji}\,\delta^{BD}\,\partial^A(\partial_A\,\underset{r}{\psi}{}^i_D - \partial_D\,\underset{r}{\psi}{}^i_A) - \frac{\mu}{2s_1}\,C^{BF}_{jk}\,\underset{r}{\psi}{}^k_F = \frac{\mu}{2s_1}\,\underset{r}{Z}{}^B_j\ ,$$

with the boundary conditions

(4.7.43) $$N_B\,C^{BF}_{jk}\,\psi^k_F\Big|_{\partial B} = -\,N_B\,Z^B_j\Big|_{\partial B}\ ,$$

(4.7.44) $$N^A(\partial_A\,\underset{r}{\psi}{}^i_B - \partial_B\,\underset{r}{\psi}{}^i_A)\Big|_{\partial B} = 0\ .$$

Again, we note that any nine functions $(\underset{r}{\psi}{}^i_A)$ that satisfy the nine field equations (4.7.42) and the $6 + 3$ boundary conditions (4.7.43) and (4.7.44) will satisfy the field equations (4.7.41) identically throughout the body. The r-th order problem is thus given by

$$\delta_{ji}\,\delta^{BD}\,\partial^A(\partial_A\,\underset{r}{\psi}{}^i_D - \partial_D\,\underset{r}{\psi}{}^i_A) - \frac{\mu}{2s_1}\,C^{BF}_{jk}\,\underset{r}{\psi}{}^k = \frac{\mu}{2s_1}\,\underset{r}{Z}{}^B_j\ ,$$

$$N_B\,C^{BF}_{jk}\,\underset{r}{\psi}{}^k_F\Big|_{\partial B} = -\,N_B\,\underset{r}{Z}{}^B_j\Big|_{\partial B}\ ,$$

$$N^A(\partial_A\,\underset{r}{\psi}{}^i_B - \partial_B\,\underset{r}{\psi}{}^i_A)\Big|_{\partial B} = 0\ ,$$

where the displacement functions are specified by

$$\underset{r}{u}{}^i = H(\underset{r}{\psi}{}^i_A\,dX^A)\ .$$

This work is based in part on a thesis, "A Complete Field Theory for Continua with Dislocations and Disclinations", submitted to Lehigh University in partial fulfillment of the requirements for the Ph.D. degree.

Appendix 1: The Lie Algebra of $SO(3) \triangleright T(3)$

The generating matrices $\underset{\sim}{\gamma}_\alpha$ of the semi-simple group SO(3) are given by:

$$(1) \qquad \underset{\sim}{\gamma}_1 = \begin{pmatrix} 0 & 0 & 0 \\ 0 & 0 & -1 \\ 0 & 1 & 0 \end{pmatrix}, \quad \underset{\sim}{\gamma}_2 = \begin{pmatrix} 0 & 0 & 1 \\ 0 & 0 & 0 \\ -1 & 0 & 0 \end{pmatrix},$$

$$\underset{\sim}{\gamma}_3 = \begin{pmatrix} 0 & -1 & 0 \\ 1 & 0 & 0 \\ 0 & 0 & 0 \end{pmatrix},$$

and their commutation relations are

$$(2) \qquad [\underset{\sim}{\gamma}_\alpha, \underset{\sim}{\gamma}_\beta] \equiv \underset{\sim}{\gamma}_\alpha \underset{\sim}{\gamma}_\beta - \underset{\sim}{\gamma}_\beta \underset{\sim}{\gamma}_\alpha = C^\xi_{\alpha\beta} \underset{\sim}{\gamma}_\xi .$$

Here, $C^\xi_{\alpha\beta}$ are the structure constants of the group SO(3). They satisfy the anti-symmetry conditions

$$(3) \qquad C^\xi_{\alpha\beta} = - C^\xi_{\beta\alpha}$$

and the Jacobi identities

$$(4) \qquad C^\delta_{\alpha\beta} C^\varepsilon_{\delta\gamma} + C^\delta_{\beta\gamma} C^\varepsilon_{\delta\alpha} + C^\delta_{\gamma\alpha} C^\varepsilon_{\delta\beta} = 0 .$$

The non-zero structure constants are determined by (3) and

$$(5) \qquad C^1_{23} = 1, \quad C^2_{31} = 1, \quad C^3_{12} = 1 .$$

The components of the Cartan-Killing metric are defined in terms of the structure constants by

(6) $$C_{\alpha\beta} = C^{\gamma}_{\alpha\delta}\, C^{\delta}_{\beta\gamma} = C_{\beta\alpha}$$

and, for the rotation group SO(3) , we have

(7) $$C_{\alpha\beta} = \delta_{\alpha\beta} \;.$$

Also, since the Cartan-Killing metric is nonsingular for semi simple groups, it has an inverse, $C^{\beta\gamma}$ given by

(8) $$C^{\beta\gamma} = \delta^{\beta\gamma} \;.$$

The generating matrices $\underset{\sim}{t}_i$ of the translation group T(3) are given by

(9) $$\underset{\sim}{t}_1 = \begin{Bmatrix} 1 \\ 0 \\ 0 \end{Bmatrix} , \quad \underset{\sim}{t}_2 = \begin{Bmatrix} 0 \\ 1 \\ 0 \end{Bmatrix} , \quad \underset{\sim}{t}_3 = \begin{Bmatrix} 0 \\ 0 \\ 1 \end{Bmatrix} .$$

The multiplication laws for the generators $\underset{\sim}{\gamma}_\alpha$ and $\underset{\sim}{t}_i$ are thus given by

(10) $$\underset{\sim}{\gamma}_\alpha \underset{\sim}{t}_i = m^j_{\alpha i}\, \underset{\sim}{t}_j \;.$$

The constants $m^j_{\alpha i}$ are anti-symmetric in the lower two indices,

(11) $$m^j_{\alpha i} = - m^j_{i\alpha} \;,$$

and they are explicitely given by (11) and

(12) $$m^1_{23} = 1 \;, \quad m^2_{31} = 1 \;, \quad m^3_{12} = 1$$

with all other enteries zero.

Appendix 2: Invariance of L_o under SO(3)⊳T(3)

Consider the vector space V_4 and the hyperplane V of V_4 consisting of all vectors of the form

(1) $$\hat{\underset{\sim}{\chi}} = \begin{Bmatrix} \underset{\sim}{\chi} \\ 1 \end{Bmatrix} , \quad \underset{\sim}{\chi} \varepsilon V_3 .$$

Let

(2) $$\underset{\sim}{M} = \begin{pmatrix} \underset{\sim}{A} & \{\underset{\sim}{b}\} \\ [\underset{\sim}{0}] & 1 \end{pmatrix}$$

be an element of the faithful matrix representation of the group G = SO(3)⊳T(3) . The action of the group G on a state vector $\hat{\underset{\sim}{\chi}}$ is expressed by

(3) $$'\hat{\underset{\sim}{\chi}} = \underset{\sim}{M}\,\hat{\underset{\sim}{\chi}} = \begin{Bmatrix} \underset{\sim}{A}\underset{\sim}{\chi} + \underset{\sim}{b} \\ 1 \end{Bmatrix}$$

and the induced transformation of the exterior covariant derivative of the state vector $\hat{\underset{\sim}{\chi}}$ is given by

(4) $$`\hat{D}`\hat{\underset{\sim}{\chi}} = \underset{\sim}{M}\hat{D}\hat{\underset{\sim}{\chi}} ,$$

where $\hat{D} = dX^a \wedge \hat{D}_a = d + \hat{\underset{\sim}{\Gamma}}\wedge$ is exterior covariant differentiation operator with respect to the full group G = SO(3)⊳T(3) .

From (1) it follows that $\partial_a \hat{\underset{\sim}{\chi}} = \left\{ \begin{matrix} \partial_a \underset{\sim}{\chi} \\ 0 \end{matrix} \right\}$ and

$$
\begin{aligned}
\hat{C}_{ab} &= \partial_a \hat{\underset{\sim}{\chi}}^T \, \partial_b \hat{\underset{\sim}{\chi}} = [\partial_a \underset{\sim}{\chi}^T, \ 0] \left\{ \begin{matrix} \partial_b \underset{\sim}{\chi} \\ 0 \end{matrix} \right\} \\[2mm]
&= \partial_a \underset{\sim}{\chi}^T \, \partial_b \underset{\sim}{\chi} = C_{ab}
\end{aligned}
$$

(5)

is the Cauchy strain tensor. Under the homogeneous action of the group G, namely under the transformation (3) of the state vector $\hat{\underset{\sim}{\chi}}$ with $d\underset{\sim}{A} = \underset{\sim}{0}$, $d\underset{\sim}{b} = \underset{\sim}{0}$, and $\underset{\sim}{A}^T \underset{\sim}{A} = \underset{\sim}{I}$, the Cauchy strain tensor is invariant and therefore any function of its elements is invariant. In particular, the Lagrangian $L_o = L_o(C_{ab})$ for elasticity theory is invariant. We then proceed to show that this is true also under the inhomogeneous action of the full group G. If we allow the matrix $\underset{\sim}{M}$ to become coordinate dependent, so that $d\underset{\sim}{A} \neq \underset{\sim}{0}$ and $d\underset{\sim}{b} \neq \underset{\sim}{0}$, we can obtain that G is an invariance group of the Lagrangian L_o.

In defect dynamics the deformation gradients $\partial_a \hat{\underset{\sim}{\chi}}$ are replaced by the distortions

(6)
$$
\hat{\underset{\sim}{B}}_a = \partial_a \hat{\underset{\sim}{\chi}} + \hat{\underset{\sim}{\Gamma}}_a \hat{\underset{\sim}{\chi}} = \left\{ \begin{matrix} \partial_a \underset{\sim}{\chi} + \underset{\sim}{\Gamma}_a \underset{\sim}{\chi} + \underset{\sim}{\phi} \\ 0 \end{matrix} \right\} = \left\{ \begin{matrix} \underset{\sim}{B}_a \\ 0 \end{matrix} \right\}
$$

where we have used the minimal replacement argument for

the group G . Under the action of G , the distortions transform as follows

$$(7) \qquad '\hat{\underset{\sim}{B}} = \underset{\sim}{M}\,\hat{\underset{\sim}{B}} = \begin{pmatrix} \underset{\sim}{A} & \{\underset{\sim}{b}\} \\ [0] & 1 \end{pmatrix} \begin{Bmatrix} \underset{\sim}{B}_a \\ 0 \end{Bmatrix} .$$

According to (5) and (6) after breaking the homogeneity of the group G_o we can write

$$(8) \qquad \hat{C}_{ab} = \hat{\underset{\sim}{B}}{}^T_a\,\hat{\underset{\sim}{B}}_b = \underset{\sim}{B}{}^T_a\,\underset{\sim}{B}_b = C_{ab} .$$

Under the inhomogeneous action of the group $SO(3) \triangleright T(3)$ the Cauchy strain tensor transforms according to

$$'C_{ab} = {}'\hat{C}_{ab} = {}'\hat{\underset{\sim}{B}}{}^T_a\,{}'\hat{\underset{\sim}{B}}_b = \hat{\underset{\sim}{B}}{}^T_a\,\underset{\sim}{M}^T\,\underset{\sim}{M}\,\hat{\underset{\sim}{B}}_b$$

which after matrix multiplications gives

$$(9) \qquad '\hat{C}_{ab} = [\underset{\sim}{B}{}^T_a,\ 0] \begin{Bmatrix} \underset{\sim}{A}^T\,\underset{\sim}{A}\,\underset{\sim}{B}_b \\ \underset{\sim}{b}^T\,\underset{\sim}{A}\,\underset{\sim}{B}_b \end{Bmatrix} = \underset{\sim}{B}{}^T_a\,\underset{\sim}{B}_b = \hat{\underset{\sim}{B}}{}^T_a\,\hat{\underset{\sim}{B}}_b = \hat{C}_{ab}$$

Hence, from (8) and (9) it follows that the Lagrangian $L_o = L_o(C_{ab})$ is invariant

$$L_o('C_{ab}) = L_o(C_{ab})$$

under the inhomogeneous action of the group G .

Appendix 3: Induced Transformations of the Field Variables

Let $\underset{\sim}{Z}$ be a row matrix whose entries $\{Z_i\}$ are defined by (3.9.25) and (3.9.29). Having in mind that the derivative of a scalar function with respect to a column matrix gives a row matrix we can symbolicly write (3.9.25) as follows

$$(1) \qquad \underset{\sim}{Z} = \frac{\partial L}{\partial \underset{\sim}{B}} .$$

Under the action of the group G the column matrix $\underset{\sim}{B}$ transforms by

$$(2) \qquad '\underset{\sim}{B} = \underset{\sim}{A}\underset{\sim}{B} , \quad \underset{\sim}{A} \epsilon G .$$

Let, now

$$(3) \qquad '\underset{\sim}{Z} = \frac{\partial L('\underset{\sim}{B}, '\underset{\sim}{\mathcal{D}}, '\underset{\sim}{\Theta})}{\partial '\underset{\sim}{B}}$$

Then, due to the invariance of the Lagrangian L under the action of G we can write $L('\underset{\sim}{B}, '\underset{\sim}{\mathcal{D}}, '\underset{\sim}{\Theta}) = L(\underset{\sim}{B}, \underset{\sim}{\mathcal{D}}, \underset{\sim}{\Theta})$ and hence

$$(4) \qquad '\underset{\sim}{Z} = \frac{\partial L(\underset{\sim}{B}, \underset{\sim}{\mathcal{D}}, \underset{\sim}{\Theta})}{\partial '\underset{\sim}{B}} = \frac{\partial L(\underset{\sim}{B}, \underset{\sim}{\mathcal{D}}, \underset{\sim}{\Theta})}{\partial \underset{\sim}{B}} \frac{\partial \underset{\sim}{B}}{\partial '\underset{\sim}{B}}$$

By (1) and (2) it follows immediately that the transformation law of the row matrix $\underset{\sim}{Z}$ is given by:

$$(5) \qquad '\underset{\sim}{Z} = \underset{\sim}{Z}\underset{\sim}{A}^{-1} .$$

Similarly, if we set

$$(6) \qquad \underset{\sim}{R} = \frac{\partial L}{\partial \underset{\sim}{\mathcal{D}}}$$

and if we use the transformation law of the column matrix

$$(7) \qquad '\underset{\sim}{\mathcal{D}} = \underset{\sim}{A}\underset{\sim}{\mathcal{D}}$$

then the row matrix $\underset{\sim}{R}$ transforms according to

$$(8) \qquad '\underset{\sim}{R} = \underset{\sim}{R}\underset{\sim}{A}^{-1} .$$

We proceed to prove the transformation law of the square matrix $\underset{\sim}{G}$ whose entries are $\{C^{\alpha\beta}\ G_\beta\}$. Under the action of the group G , the curvature matrix $\underset{\sim}{\Theta} = F^\alpha\ \underset{\sim}{\gamma}_\alpha$ transforms according to

$$(9) \qquad '\underset{\sim}{\Theta} = \underset{\sim}{A}\underset{\sim}{\Theta}\underset{\sim}{A}^{-1} .$$

Then, if we define the orthogonal matrix H^α_β by

$$(10) \qquad H^\alpha_\beta\ \underset{\sim}{\gamma}_\alpha = \underset{\sim}{A}\underset{\sim}{\gamma}_\beta\underset{\sim}{A}^{-1} ,$$

we can write the transformation law for the 2-forms F^α

$$(11) \qquad 'F^\alpha = F^\beta\ H^\alpha_\beta .$$

The transformation law of the Cartan-Killing metric, $C_{\alpha\beta}$ is

$$(12) \qquad C_{\delta\gamma} = 'C_{\alpha\beta}H^\alpha_\gamma\ H^\beta_\delta ,$$

which is essential in establishing invariance of the Lagrangian L . Accordingly, the transformation law of the inverse Cartan-Killing metric is given by

$$(13) \qquad {}'C^{\alpha\beta} = C^{\gamma\delta} H^{\alpha}_{\gamma} H^{\beta}_{\delta} .$$

From (3.14.8), (11) and the invariance of the Largangian L we obtain

$$(14) \qquad {}'G_{\alpha} = G_{\beta} \overset{-1}{H}{}^{\beta}_{\alpha}$$

Hence, (10), (13) and (14) give

$$(15) \qquad {}'\underset{\sim}{G} = {}'C^{\alpha\beta}\, {}'G_{\beta} \underset{\sim}{Y}_{\alpha} = C^{\gamma\delta} G_{\mu} H^{\beta}_{\delta} \overset{-1}{H}{}^{\mu}_{\beta} H^{\alpha}_{\gamma} \underset{\sim}{Y}_{\alpha}$$

$$= G^{\gamma} H^{\alpha}_{\gamma} \underset{\sim}{Y}_{\alpha} = \underset{\sim}{A} G^{\delta} \underset{\sim}{Y}_{\delta} \underset{\sim}{A}^{-1} = \underset{\sim}{A}\underset{\sim}{G}\underset{\sim}{A}^{-1} ,$$

and the transformation law for the square matrix $\underset{\sim}{G}$ is established:

$${}'\underset{\sim}{G} = \underset{\sim}{A}\underset{\sim}{G}\underset{\sim}{A}^{-1} .$$

Appendix 4: A Four-Dimensional Formulation of Defects - Dynamics and Thermodynamics

1. Introduction

Recent studies of the equations of defect dynamics [33,37,38] reveal a marked difference between these field equations and the field equations of classical continuum theories, for the field equations of defect dynamics admit nontrivial Abelian gauge groups while the equations of classical continuum theories do not. The Abelian gauge groups of defect dynamics are not only non-trivial, they are also much richer than the analogous gauge group of classical electrodynamics since it has been shown that there are at least 27 independent generators [38]. This richness arises through the interplay of three distinct constructs: the geometric response of a body to systems of loadings, the evolution of dislocations within the body, and evolution of disclinations within the body. Unfortunately, the customary 3-dimensional space plus time formulation of defect dynamics does not provide a clear separation of these three essentially different aspects of the theory. A similar situation arises in electrodynamics which the 3-dimensional space plus time formulation yields field equations that are admixtures of effects. Since electrodynamics is significantly simplified and systematized by a reformulation in a 4-dimensional space-time setting, an argument by analogy suggests that a reformulation of the equations of defect dynamics in a 4-dimensional space-time setting would also provide simplifications and disentanglements of the various structures that are involved.

2. The Field Equations of Defect Dynamics

Let E_3 denote the Euclidean 3-dimensional space with Cartesian coordinates $\{X^1, X^2, X^3\}$ that contains the reference configuration of the material body under study. The volume element of E_3 is denoted by

(2.1) $$\mu = dX^1 \wedge dX^2 \wedge dX^3 ,$$

which is also a basis for the 1-dimensional vector space of exterior forms of degree 3 on E_3. The natural basis for the tangent space of E_3 is denoted by $\{\partial_A\} = \{\partial_1, \partial_2, \partial_3\}$, where $\partial_B f = \partial f/\partial X^B$, and induces a basis $\{\mu_A\}$ (see [3,38])

(2.2) $$\mu_A = \partial_A \rfloor \mu ,$$

for exterior forms of degree 2 on E_3. These base elements have the property that

(2.3) $$d\mu_A = 0 , \quad dX^A \wedge \mu_B = \delta^A_B \, \mu$$

and may be viewed as oriented two-dimensional surface elements.

The reason for laboring the notation through the formalism of the exterior calculus is that it is a natural vehicle for the exposition of both the field theory of defect dynamics and electrodynamics. To be more specific, we define the following defect dynamic fields of exterior forms:

$\alpha^i = \alpha^{Ai}\mu_A$ = 2-forms of dislocation density ,

$J^i = J^i_A dX^A$ = 1-forms of dislocation current ,

$S^i = S^{Ai}\mu_A$ = 2-forms of disclination current ,

$\Theta^i = \theta^i\mu$ = 3-forms of disclination density .

In general, all quantities are functions of time, T , as well as the three spatial variables X^1, X^2, X^3 . If we use $\bar{d}$ to denote the operation of exterior differentiation with T held fixed and ∂_4 to denote differentiation with respect to the time variable, T , the continuity equations of defect dynamics assume the simple forms

$$\partial_4\alpha^i = -\bar{d}J^i - S^i \quad , \quad \partial_4\Theta^i = -\bar{d}S^i \quad , \tag{2.4}$$

$$\bar{d}\alpha^i = \Theta^i \quad , \quad \bar{d}\Theta^i = 0 \quad . \tag{2.5}$$

When these equations are resolved on the basis elements $\{\mu, \mu_A\}$, they assume the equivalent forms familiar in the literature [10,17,19]:

$$\partial_4\alpha^{Ai} = -S^{Ai} - e^{ABC}\partial_B J^i_C \quad , \quad \partial_4\theta^i = -\partial_A S^{Ai} \; ,$$

$$\partial_A\alpha^{Ai} = \theta^i \; .$$

(Note that there is no resolution of the second of (2.5) since a 4-form on a 3-dimensional space vanishes identically).

In a like manner, we introduce the remaining fields of defect dynamics through the definitions

$\beta^i = \beta^i dX^A$ = 1-forms of distortion ,

$\omega^i = \omega^i_A dX^A$ = 1-forms of spin ,

$k^i = K^{Ai}\mu_A$ = 2-forms of bend-twist ,

V^i = 0-forms of velocity.

The defect field equations are given in this notation in the following form:

$$(2.6) \qquad \partial_4 k^i = \bar{d}\omega^i - S^i \qquad \bar{d}k^i = \theta^i ,$$

$$(2.7) \qquad \partial_4 \beta^i = \bar{d}V^i - J^i - \omega^i , \quad \bar{d}\beta^i = \alpha^i - k^i .$$

Again, a resolution on the basis elements $\{\mu_A, dX^A\}$ gives the field equations in their possibly more familiar component forms:

$$\partial_4 k^{Ai} = - S^{Ai} + e^{ABC}\partial_B \omega^i_C \quad , \quad \partial_A k^{Ai} = \theta^i ,$$

$$\partial_4 \beta^i_A = \partial_A V^i - J^i_A - \omega^i_A \quad , \quad e^{ABC}\partial_B \beta^i_C = \alpha^{Ai} - k^{Ai} .$$

The field equations listed thus far primarily describe the internal state of a body with defects; in effect, they replace the more classic equations of the kinematics of continuum ($\bar{d}\beta^i = 0$, $\partial_4 \beta^i = \bar{d}V^i$) that imply $\beta^i_A = \partial_A x^i$, $V^i = \partial_4 x^i$, $x^i = \chi^i(X^A,T)$ = coordinates of particle in the current configuration that had coordinates X^A in the reference configuration. These equations must be supplemented by appropriate statements of the laws of balance of linear momentum and energy before a complete theory can be obtained.

Let

(2.8) $\sigma_i = \sigma_i^A \mu_A$ = 2-forms of mechanical surface traction,

(2.9) $P_i = p_i \mu$ = 3-forms of momentum density ,

(2.10) $W = W^A \mu_A$ = 2-form of total energy flux ,

(2.11) $E = e\mu$ = 3-form of total energy density .

The coefficients $(\sigma_i^A , p_i , W^A , e)$ are assumed to be given in terms of a comprehensive system of constitutive relations, the restrictions on which will be discussed later. For the time being, we need only consider the exterior differential equations

(2.12) $\partial_4 P_i = \bar{d}\sigma_i$,

(2.13) $\partial_4 E = \bar{d}W$.

When these equations are resolved on the base elements $\{\mu , \mu_A\}$ and we use the relations $\bar{d}(Y^A \mu_A) = \partial_A Y^A \mu$ that follow directly from (2.3), we obtain the equations

(2.14) $\partial_4 p_i = \partial_A \sigma_i^A$,

(2.15) $\partial_4 e = \partial_A W^A$.

It is thus evident that (2.12) and (2.13) do represent the basic balance statements of linear momentum and total energy. The important thing that needs to be noted is that the quantities $\{\sigma_i^A\}$ play the role of the Piola-

Kirchhoff stress, as indeed they must in view of the fact that our formulation uses the reference configuration coordinates as the independent spatial variables.

In order to summarize the equations of defect dynamics, we need only note that the system (2.6), (2.7) constitute first integrals of the system (2.4), (2.5). We therefore have the system of field equations

$$\partial_4 P_i = \bar{d}\sigma_i \quad , \quad \partial_4 E = \bar{d}W \; , \tag{2.16}$$

and

$$\partial_4 \alpha^i = \bar{d}J^i - S^i \quad , \quad \partial_4 \theta^i = -\bar{d}S^i \; , \tag{2.17}$$

$$\bar{d}\alpha^i = \theta^i \quad , \quad \bar{d}\theta^i = 0 \; , \tag{2.18}$$

together with the first integrals

$$\partial_4 k^i = \bar{d}\omega^i - S^i \quad , \quad \bar{d}k^i = \theta^i \; , \tag{2.19}$$

$$\partial_4 \beta^i = \bar{d}V^i - J^i - \omega^i \quad , \quad \bar{d}\beta^i = \alpha^i - k^i \; . \tag{2.20}$$

3. Preliminaries to A 4-Dimensional Formulation

The formulation of the equations of defect dynamics given in the last section was based on the assumptions of a three-dimensional Euclidean reference space, E_3 , and an independent time scale that was taken to be the real line. Although this space + time structure is an adequate basis for the construction of continuum theories and theories of materials with defects, it does not reveal certain intrinsic structural relations and analogies as readily as a truly four-dimensional space-time formulation. Noting that the equations of defect dynamics are given in terms of exterior forms, the translation of these equations into a fully four-dimensional setting becomes almost immediate once the relations between the three-dimensional and the four-dimensional basis elements for exterior forms of various degrees have been obtained.

Let E_4 denote the four-dimensional reference configuration with the Cartesian coordinate cover $\{X^a\} = \{X^1, X^2, X^3, X^4\}$. We consider the four element index set $\{a\}$ as partitioned into a three element index set $\{A\}$ and a one element index set consisting of the element 4 ; $\{a\} = \{A,4\}$. Let us further agree to set $X^4 = T$ whenever we replace $\{a\}$ by $\{A,4\}$, so that we may write $\{X^a\} = \{X^A,T\}$.

The volume element of E_4 , which is also a basis for 4-forms, is given by

$$(3.1)\qquad \omega = dX^1 \wedge dX^2 \wedge dX^3 \wedge dX^4 = \frac{1}{4!} e_{abef} dX^a \wedge dX^b \wedge dX^e \wedge dX^f .$$

Thus, since $\mu = dX^1 \wedge dX^2 \wedge dX^3$, we obtain the elementary result

$$(3.2)\qquad \omega = \mu \wedge dT .$$

The natural basis for the tangent space of E_4 is given by the four linear operators $\{\partial_a, a=1,\ldots,4\}$ with $\partial_a f = \partial f/\partial X^a$, and a basis for 3-forms is given by

$$(3.3)\qquad \pi_a = \partial_a \lrcorner\, \omega = \frac{1}{3!} e_{abef} dX^b \wedge dX^e \wedge dX^f , \quad a=1,2,3,4 .$$

Here $\lrcorner$ is used to denote the operation of inner multiplication [3, Appendix]. These base elements exhibit the properties

$$(3.4)\qquad d\pi_a = 0 , \quad dX^a \wedge \pi_b = \delta^a_b \omega ,$$

where $d = dX^a \wedge \partial_a$ denotes the 4-dimensional exterior derivative operator.

We now need to relate these quantities to those that occur in the space + time formulation. It is clear from the definition of the operator d that $d = dX^a \wedge \partial_a$ $= dX^A \wedge \partial_A + dT \wedge \partial_4$, while $\bar{d} = dX^A \wedge \partial_A$. We thus have

(3.5) $$d = \bar{d} + dT \wedge \partial_4$$

It is now a simple matter to substitute (3.2) into (3.3) to obtain $\pi_a = (\partial_a \lrcorner \mu) \wedge dT - \mu \wedge (\partial_a \lrcorner dT)$. However, $\partial_4 \lrcorner \mu = 0$, $\partial_4 \lrcorner dT = 1$, $\partial_A \lrcorner dT = 0$, $\partial_A \lrcorner \mu = \mu_A$, and hence

(3.6) $$\pi_a = \delta^A_a \mu_A \wedge dT - \delta^4_a \mu .$$

Since any 3-form on E is uniquely expressible in the form

$$\Gamma = \Gamma^a \pi_a = \Gamma^A \pi_A + \Gamma^4 \pi_4 ,$$

use of (3.6) provides the following conclusion:

Any 3-form Γ *on* E_4 *is uniquely determined by the 2-form* $\Gamma^A \mu_A$ *and the 3-form* $\gamma\mu = -\Gamma^4 \mu$ *through the relation*

(3.7) $$\Gamma = \Gamma^A \mu_A \wedge dT + \gamma\mu .$$

A direct combination of (3.4), (3.5) and (3.7) yields the following complementary conclusion:

Any 3-form Γ *on* E_4 *that is given by (3.7) has its exterior derivative determined in terms of the 3-forms* $\bar{d}(\Gamma^A \mu_A)$ *and* $\partial_4 \Gamma^4 \mu$ *by*

(3.8) $$d\Gamma = (\bar{d}(\Gamma^A \mu_A) - \partial_4 \gamma\mu) \wedge dT .$$

A basis for 2-forms on E is given by

(3.9) $$\pi_{ab} = \partial_a \lrcorner \pi_b = \partial_a \lrcorner (\partial_b \lrcorner \omega) ,$$

and they have the following properties:

(3.10) $$\pi_{ab} = -\pi_{ba}\ , \quad d\pi_{ab} = 0\ ,$$

$$dX^c \wedge \pi_{ab} = \delta^c_a \mu_b - \delta^c_b \pi_a\ .$$

If Σ is a 2-form on E_4, it can be expressed uniquely in the form $\Sigma = \Sigma^{ab}\ \pi_{ab}$ with $\Sigma^{ab} = -\Sigma^{ba}$. Application of the same procedure as that used for 3-forms then gives the following results.

A 2-form Σ *on* E_4 *is uniquely determined by a 1-form* $\rho = \rho_A dX^A$ *and a 2-form* $\eta = \eta^A\ \mu_A$ *through the relation*

(3.11) $$\Sigma = \rho \wedge dT + \eta = \rho_A dX^A \wedge dT + \eta^A \mu_A$$

and its exterior derivative is given by

(3.12) $$d\Sigma = (\bar{d}\rho + \partial_4 \eta) \wedge dT + \bar{d}\eta$$

$$= (\bar{d}(\rho_A dX^A) + \partial_4 \eta^A \mu_A) \wedge dT + \bar{d}(\eta^A \mu_A)\ .$$

4. Defect Dynamics in 4-Dimensional Space-Time

The transition to a full 4-dimensional description of defect dynamics is now simply a matter of comparing the forms of the results (3.7), (3.8), (3.11) and (3.12) with the field equations (2.16) through (2.20). Let us start with (2.16). Since $\sigma_i = \sigma_i^A \mu_A$, $W = W^A \mu$ are 2-forms of the appropriate type (i.e., are 2-forms on E_3 with coefficients that depend on the location in E_4), while $P_i = P_i \mu$, $E = e \mu$ are 3-forms of the appropriate type, we may use them to define the *3-forms of stress-momentum*

$$\mathcal{P}_i = \sigma_i \wedge dT + P_i = \sigma_i^A \mu_A \wedge dT + p_i \mu$$

and the *3-form of energetics*

$$\mathcal{P}_4 = W \wedge dT + E = W^A \mu_A \wedge dT + e \mu .$$

Since the exterior derivatives of these 3-forms give

$$d\mathcal{P}_i = (\bar{d}\sigma_i - \partial_4 P_i) \wedge dT , \quad d\mathcal{P}_4 = (\bar{d}W - \partial_4 E) \wedge dT ,$$

a comparison with (2.16) yields the following conclusions:

The laws of balance of linear momentum and of energy are satisfied if and only if the 3-forms

$$\mathcal{P}_i = \sigma_i \wedge dT + P_i = \sigma_i^A \mu_A \wedge dT + p_i \mu \tag{4.1}$$

of stress-momentum and the 3-form

(4.2) $$P_4 = W \wedge dT + E = W^A \mu_A \wedge dT + e\,\mu$$

of energetics satisfy the exterior equations

(4.3) $$dP_i = 0 \;,\quad dP_4 = 0$$

throughout the region of E_4 *that is occupied by the space-time history of the body.*

In the case of (2.17) and (2.18), we first consider the *disclination 3-forms*

$$\Omega^i = -\, S^i \wedge dT + \Theta^i = -\, S^{Ai} \mu_A \wedge dT + \theta^i \mu \;.$$

Since these 3-forms have the structure given by (3.7), (3.8) yields

$$d\Omega^i = -\, (\bar{d}S^i + \partial_4 \Theta^i) \wedge dT \;.$$

This shows that the second of (2.17) is satisfied if and only if $d\Omega^i = 0$. Further, since $\Theta^i = \theta^i \mu$, we have that $d\Theta^i = d\theta^i \wedge \mu = \partial_A \theta^i \, dX^A \wedge \mu$ vanishes identically since $dX^A \wedge \mu$ vanishes identically. The second of (2.18) is thus identically satisfied because $\Theta^i = \theta^i \mu$. In order to obtain the first of (2.17) and of (2.18), we construct the *dislocation 2-forms*

$$\mathcal{D}^i = J^i \wedge dT + \alpha^i = J^i_A \, dX^A \wedge dT + \alpha^{Ai} \mu_A \;.$$

Since these are of the form given by (3.11), (3.12) gives

$$d\mathcal{D}^i = (\bar{d}J + \partial_4 \alpha^i) \wedge dT + \bar{d}\alpha^i \;.$$

When the first of (2.17) and (2.18) are used to eliminate $\bar{d}J + \partial_4 \alpha^i$ and $\bar{d}\alpha^i$, we obtain

$$d\mathcal{D}^i = - S^i \wedge dT + \Theta^i .$$

Thus, (2.17) and (2.18) are satisfied if and only if $\mathcal{D}^i$ and Ω^i stand in the relation $d\mathcal{D}^i = \Omega^i$.

The system of field equations (2.17), (2.18) *of defect dynamics are satisfied if and only if the disclination 3-forms*

$$(4.4) \qquad \Omega^i = - S^i \wedge dT + \Theta^i = - S^{Ai} \mu_A \wedge dT + \theta^i \mu$$

and the dislocation 2-forms

$$(4.5) \qquad \mathcal{D}^i = J^i \wedge dT + \alpha^i = J^i_A \, dX^A \wedge dT + \alpha^{Ai} \mu_A$$

satisfy the exterior differential equations

$$(4.6) \qquad d\Omega^i = 0 \; , \quad d\mathcal{D}^i = \Omega^i .$$

throughout the region of E_4 *that is occupied by the space-time history of the body.*

A useful insight already emerges from the 4-dimensional formulation, for (4.4) and (4.5) show that the dislocation related quantities occur only through the 2-forms $\mathcal{D}^i$ while the disclination related quantities enter only by way of the 3-forms Ω^i . There is thus a specific and irremovable distinction between dislocation related and disclination related structures that is not necessarily evident from the 3-dimensional formulation!

It might be expected that the first integrals, namely (2.19) and (2.20), of the exterior system (2.17), (2.18) may likewise be obtained. This is indeed the case as we now proceed to show. If we solve (2.20) for J^i and α^i, $J^i = \bar{d}V^i - \omega^i - \partial_4\beta^i$, $\alpha^i = \bar{d}\beta^i + k^i$, and substitute these into (4.5), we obtain

$$\mathcal{D}^i = (-\omega^i \wedge dT + k^i) + (\bar{d}V^i - \partial_4\beta^i) \wedge dT + \bar{d}\beta^i \; . \tag{4.7}$$

Thus, if we define the *velocity-distortion* 1-forms by

$$B^i = V^i\, dT + \beta^i = V^i\, dT + \beta^i_A\, dX^A \tag{4.8}$$

and the *spin-twist* 2-forms through

$$K^i = -\,\omega^i \wedge dT + k^i = -\,\omega^i_A\, dX^A \wedge dT + k^{Ai}\mu_A \; , \tag{4.9}$$

the system (2.20) is satisfied as a consequence of the exterior differential relations

$$\mathcal{D}^i = dB^i + K^i \; . \tag{4.10}$$

When the exterior derivative of (4.10) is taken, we obtain the relations $d\mathcal{D}^i = dK^i$. A combination of these with $d\mathcal{D}^i = \Omega^i$ then leads to the satisfaction of the system (2.19).

The system of field equations

(4.11) $$d\mathcal{D}^i = \Omega^i$$

admits the system of first integrals

(4.12) $$\mathcal{D}^i = d\mathcal{B}^i + \mathcal{K}^i$$

in terms of the velocity-distortion 1-*forms*

(4.13) $$\mathcal{B}^i = V^i\, dT + \beta^i = V^i\, dT + \beta^i_A\, dX^A$$

and the spin-twist 2-*forms*

(4.14) $$\mathcal{K}^i = -\,\omega^i \wedge dT + k^i = -\,\omega^i_A\, dX^A \wedge dT + k^{Ai}\,\mu_A\ .$$

Again, we have a useful separation of effects, velocity and distortion on the one hand versus spin and bend-twist on the other hand, that comes about through the difference of the degrees of the exterior forms $\mathcal{B}^i$ and $\mathcal{K}^i$.

5. The 45-Fold Abelian Gauge Group and the Indeterminism Question

An argument from analogy with the field equations of electrodynamics was used by A. A. Golebiewska-Lasota in [33] to show that the equations of dislocation dynamics admitted a nontrivial system of Abelian gauge transformations. The conceptual disentanglements that resulted from exploitation of these gauge transformations were such that an analysis of the gauge transformations for defect dynamics with linear strain measures was given in [37] and the general case was considered in [38]. The latter established the existence of a 27-fold Abelian gauge group with the property that α^i and Θ^i were left invariant by J^i , S^i , ω^i , k^i and β^i were changed in an additive way. Certain specific choices of the 27 functions that generated the changes were shown to lead to a formulation in which the various responses of the body could be represented in terms of sums of internal modes of response and external modes of response. Once the internal and external modes of response were identified, an analysis by means of the practices of irreversible thermodynamics become a manageable task and lead to a well defined procedure for the specification of the necessary constitutive relations of the theory. An analysis of the gauge transformations of the 4-dimensional formulation is thus required in order that similar results may be obtained.

The basic field equations for the defects are given by

(5.1) $d\Omega^i = 0 \ , \quad \Omega^i = d\mathcal{D}^i \ ,$

(5.2) $\mathcal{D}^i = d\mathcal{B}^i + \mathcal{K}^i \ , \quad d\mathcal{K}^i = \Omega^i \ .$

The underlying idea of a gauge group (more precisely, an Abelian gauge group) is that the field variables, say f^α, are set equal to $\bar{f}^\alpha + g^\alpha$ and the gauge generators, g^α, are chosen so as to force the new field variables, $\bar{f}^\alpha$, to satisfy the same field equations as were satisfied by the f^α's . This is a straightforward calculation that parallels that given in [38], so we shall simply state the results.

Each choice of the 45 functions ν^i_A , ν^{Ai} , f^i , F^i_A , g^i , G^i_A , χ^i *defines a system of 2-forms*

(5.3) $\nu^i = \nu^i_a \, dX^A \wedge dT + \nu^{Ai} \, \mu_A \ ,$

a system of 1-forms

(5.4) $F^i = f^i \, dT + F^i_A \, dX^A \ , \quad G^i = g^i \, dT + G^i_A \, dX^A \ ,$

and a system of 0-forms χ^i *such that the transformations*

(5.5) $\Omega^i = \bar{\Omega}^i + d\nu^i \ ,$

(5.6) $\mathcal{D}^i = \bar{\mathcal{D}}^i + \nu^i + dF^i \ ,$

$$(5.7) \qquad B^i = \bar{B}^i + d\chi^i + F^i - G^i ,$$

$$(5.8) \qquad K^i = \bar{K}^i + \nu^i + dG^i$$

map the field equations (5.1), (5.2) *onto the field equations*

$$(5.9) \qquad d\bar{\Omega}^i = 0 , \quad \bar{\Omega}^i = d\bar{D}^i ,$$

$$(5.10) \qquad \bar{D}^i = d\bar{B}^i + \bar{K}^i , \quad d\bar{K}^i = \bar{\Omega}^i .$$

The field equations (5.1), (5.2) *thus admit a 45-fold gauge group.*

The first thing to be noted is that this gauge group reduces to the 27-fold gauge group reported in [38] if we require α^i and Θ^i to be invariant (i.e., if we require $\nu^{Ai} = 0$, $F^i_A = 0$, see (4.4) and (4.5)). All of the results obtained in [38] are thus directly applicable in the 4-dimensional formulation.

Second, it is necessary to realize that the presence of such a gauge group signifies that the field equations (5.1) and (5.2) have built-in indeterminisms that can lead to spurious conclusions. The clearest way of seeing this is to ask whether the field equations can describe defects that are not actually there. To this end, let us consider a body without defects. Such a situation is described by an integrable response $x^i = \psi^i(X^A,T)$, where $\{x^i\}$ are the coordinates of a material point in the current configuration that occupied the point with coordinates $\{X^A\}$ in the reference configuration. In this case, we have

$$(5.11) \qquad \mathcal{B}^i = d\psi^i \;, \quad \mathcal{D}^i = 0 \;, \quad \mathcal{K}^i = 0 \;, \quad \Omega^i = 0 \;.$$

When these are put into (5.5)-(5.8), we obtain

$$(5.12) \qquad \bar{\Omega}^i = - d\nu^i \;, \; \bar{\mathcal{D}}^i = - dF^i - \nu^i \;, \; \bar{\mathcal{K}}^i = - \nu^i - dG^i$$

and

$$(5.13) \qquad \bar{\mathcal{B}}^i = d(\psi^i - \chi^i) - F^i + G^i \;.$$

For each assignment of the forms ν^i, F^i, G^i and χ^i we obtain explicit nontrivial expressions for $\bar{\Omega}^i$, $\bar{\mathcal{D}}^i$, $\bar{\mathcal{K}}^i$ and $\bar{\mathcal{B}}^i$ that satisfy the field equations

$$d\bar{\Omega}^i = 0 \;, \quad \bar{\Omega}^i = d\bar{\mathcal{D}}^i \;,$$

$$\bar{\mathcal{D}}^i = d\bar{\mathcal{B}}^i + \bar{\mathcal{K}}^i \;, \quad d\bar{\mathcal{K}}^i = \bar{\Omega}^i \;.$$

Thus, since the field equations are satisfied, it would be difficult to say that $\bar{\Omega}^i$, $\bar{\mathcal{D}}^i$, $\bar{\mathcal{K}}^i$ and $\bar{\mathcal{B}}^i$ do not describe a state of the body with defects if we did not already know that the body was in a state that is actually defect free. Put another way, a solution of the field equations (5.1), (5.2) gives specific functions that in no way signify whether they are barred quantities or unbarred quantities so we would not know whether spurious defects have been introduced by the 45-fold gauge group. This is because the 45-fold gauge group maps solutions of the field equations onto solutions of the field equations. In fact, it is clear that the gauge group embeds any solution of the field equations (5.1), (5.2) in a family of solutions that contains 45 arbitrary functions of position and time. It should thus be abundantly clear that a specific procedure must be given whereby we may eliminate this enormous degree of indeterminism. In essence, we must provide a specific collection of gauge conditions that serve to select physically relevant solutions. In the example given above, the conditions should preclude the possibility of calculating "null" defects; that is, defects associated with states without defects.

6. Homotopy Operators and the Gauge Conditions

The first thing we need to note in the search for effective gauge conditions is that the forms

(6.1) $$\Omega^i = -S^i \wedge dT + \Theta^i \ , \quad \mathcal{D}^i = J^i \wedge dT + \alpha^i$$

characterize the disclination and dislocation densities and currents while the forms

(6.2) $$\mathcal{K}^i = -\omega^i \wedge dT + k^i \ , \quad \mathcal{B}^i = V^i \, dT + \beta^i$$

characterize quantities that arise in response to the presence of the disclination and dislocation densities and currents. It would thus seem reasonable to expect that an effective system of gauge conditions would lead to the conditions $\bar{\Omega}^i = \Omega^i$, $\bar{\mathcal{D}}^i = \mathcal{D}^i$. On the other hand, (5.6) shows that $\mathcal{D}^i = \bar{\mathcal{D}}^i$ if and only if $\nu^i = -dF^i$, and (5.5) shows that this in turn implies $\Omega^i = \bar{\Omega}^i$. Thus, the gauge conditions

(6.3) $$\nu^i = -dF^i$$

imply

(6.4) $$\Omega^i = \bar{\Omega}^i \ , \quad \mathcal{D}^i = \bar{\mathcal{D}}^i$$

and the remaining gauge transformations become

(6.5) $$\mathcal{K}^i = \bar{\mathcal{K}}^i - d(F^i - G^i) \ , \quad \mathcal{B}^i = \bar{\mathcal{B}}^i + d\chi^i + F^i - G^i \ .$$

Thus, everything will be determined once we obtain conditions for the determination of F^i , G^i and χ^i .

The key to the situation is the fact that K^i and $\bar{K}^i$ satisfy $dK^i = \Omega^i = \bar{\Omega}^i = d\bar{K}^i$, and hence most questions will be resolved if we can compute a unique $\bar{K}^i$ in terms of a given Ω^i that will satisfy $d\bar{K}^i = \Omega^i$. This, however, is exactly what the linear homotopy operator H of the exterior calculus accomplishes (see Chapter V of the appendix of [3] for a full account of this operator). Let $\{X_o^a\} = \{X_o^A, T_o\}$ be a selected point in E_4 and construct the vector field

(6.6) $$X = (X^A - X_o^A)\partial_A + (T - T_o)\partial_4 = (X^a - X_o^a)\partial_a$$

on E_4 . If ω is a form of degree k on E_4 , we use $\tilde{\omega}(\lambda)$ to denote the 1-parameter family of k forms on E_4 that is obtained from ω by replacing the coefficient function at $\{X^a\}$ by the same coefficient functions evaluated at $X_o^a + \lambda(X^a - X_o^a)$. Thus, if $\Gamma = \gamma_b(X^a)\, dX^b$, then $\tilde{\Gamma}(\lambda) = \gamma_b(X_o^a - \lambda(X^a - X_o^a))\, dX^b$. The operator H , defined in terms of these constructs by

(6.7) $$H\langle\omega\rangle = \int_o^1 X \rfloor \tilde{\omega}(\lambda)\, \lambda^{k-1}\, d\lambda \quad ,$$

is such that it satisfies the identities

(6.8) $$\omega \equiv H\langle d\omega\rangle + dH\langle\omega\rangle \quad , \quad X \rfloor H\langle\omega\rangle \equiv 0 \quad ,$$

$$H\langle H\langle\omega\rangle\rangle \equiv 0 \ .$$

If we are given the exterior differential equation $d\omega = \alpha$, (6.8) leads directly to the solution

$$(6.9) \qquad \omega = dH\langle\omega\rangle + H\langle\alpha\rangle$$

where we are free to specify $H\langle\omega\rangle$, while $H\langle\alpha\rangle$ is that part of ω that is uniquely determined by α through $d\omega = \alpha$.

If we choose

$$(6.10) \qquad F^i = -H\langle K^i\rangle + G^i \quad , \quad X^i = H\langle B^i\rangle \ ,$$

(6.5) gives

$$K^i = \bar{K}^i + dH\langle K^i\rangle \quad , \quad B^i = \bar{B}^i + dH\langle B^i\rangle - H\langle K^i\rangle \ ,$$

that is

$$(6.11) \qquad \bar{K}^i = K^i - dH\langle K^i\rangle = H\langle dK^i\rangle = H\langle \Omega^i\rangle$$

and

$$(6.12) \qquad \bar{B}^i = B^i - dH\langle B^i\rangle + H\langle K^i\rangle = H\langle dB^i + K^i\rangle = H\langle \mathcal{D}^i\rangle \ .$$

These choices of the gauge generators thus give us the relations

$$(6.13) \qquad \Omega^i = \bar{\Omega}^i \quad , \quad \mathcal{D}^i = \bar{\mathcal{D}}^i$$

and

$$(6.14) \qquad K^i = \bar{K}^i + dH\langle K^i\rangle \quad , \quad B^i = \bar{B}^i + dH\langle B^i\rangle - H\langle K^i\rangle \ ,$$

where the quantities $\bar{K}^i$ and $\bar{B}^i$ are uniquely determined in terms of the invariant quantities Ω^i and $\mathcal{D}^i$ through

$$(6.15) \qquad \bar{K}^i = H\langle \Omega^i\rangle \quad , \quad \bar{B}^i = H\langle \mathcal{D}^i\rangle \ .$$

The choice of the gauge generating functions (gauge conditions) given above possesses the pleasant circumstance of rendering $\bar{K}^i$ and $\bar{B}^i$ uniquely determined in terms of the invariant quantities Ω^i and $\mathcal{D}^i$. It now remains to show that this selection eliminates the possibility of obtaining defects where there are actually none at all. To this end, we return to the problem in which $B^i = d\psi^i$ with $x^i = \psi^i(X^A,T)$; that is, to the classic problem of the continuum. In this case, $K^i = 0$, $\mathcal{D}^i = 0$, $\Omega^i = 0$, while the gauge conditions $\nu^i = dF^i$, $F^i = -H\langle K^i\rangle + G^i$, $\chi^i = H\langle B^i\rangle$ yield $\nu^i = -dF^i$, $F^i = G^i$, $\chi^i = H\langle d\psi^i\rangle = \psi^i - c^i$, with c^i constants. Thus, (5.5)-(5.8) yield $0 = \Omega^i = \bar{\Omega}^i$, $0 = \mathcal{D}^i = \bar{\mathcal{D}}^i$, $d\psi^i = B^i = \bar{B}^i + d\psi^i$ which implies $\bar{B}^i = 0$, and $0 = K^i = \bar{K}^i$. We also obtain $\bar{B}^i = 0$, $\bar{K}^i = 0$ directly from $\bar{K}^i = H\langle\Omega^i\rangle = 0$ and $\bar{B}^i = H\langle\mathcal{D}^i\rangle = 0$, and thus conclude that *the gauge conditions given above preclude the possibility of introducing "null defects" into the theory.*

Since the definition of the linear homotopy operator H is dependent upon the selection of the point $\{X_o^a\}$, it might seem that everything would dissolve in confusion if a different choice were made. This is not the case, as we now proceed to demonstrate. Let H_o denote the homotopy operator formed from the choice $\{X_o^a\}$ and let H_1 denote the homotopy operator based on the choice $\{X_1^a\}$. It is a straightforward matter to show that H_o

and H_1 stand in the relation

(6.16) $$H_1\langle\omega\rangle = H_o\langle\omega\rangle + \rho + d\eta$$

with

(6.17) $$\rho = - H_o\langle H_1\langle d\omega\rangle\rangle \quad , \quad \eta = H_o\langle H_1\ \omega\rangle\rangle ,$$

(see Lemma 5-7.1 of [3]). The gauge transformation equations and (6.16), (6.17) then lead directly to the following conclusion.

The gauge conditions $\nu^i = - dF^i$, $F^i = - H\langle K^i\rangle + G^i$, $\chi^i = H\langle B^i\rangle$ *and their consequences are form invariant under changes in the choice of the point* $\{X_o^a\}$; that is, if we replace H by H_1 and choose the gauge by

$$\nu_1^i = - dF_1^i \ , \ F_1^i = - H_1\langle K^i\rangle + G_1^i \ , \ \chi_1^i = H_1\langle B^i\rangle ,$$

then

$$\Omega^i = \bar{\Omega}^i \ , \ \mathcal{D}^i = \bar{\mathcal{D}}^i \ , \ \bar{K}^i = K_1^i + dH_1\langle K^i\rangle \ ,$$

$$B^i = \bar{B}_1^i + dH_1\langle B^i\rangle - H_1\langle \bar{K}^i\rangle \ , \ \bar{K}_1^i = H_1\langle\Omega^i\rangle ,$$

$$\bar{B}_1^i = H_1\langle\mathcal{D}^i\rangle .$$

Another, and perhaps more meaningful way of putting this is as follows: *a change in the choice of the point* $\{X_o^A\}$ *for the evaluation of the homotopy operator gives the relations* (6.16), (6.17) *whose structure is such that* there exists a gauge transformation that maps the gauge conditions based on one homotopy operator onto the same gauge conditions based upon the other homotopy operator.

7. Explicit Evaluations and the Plastic Velocity and Distortion

Explicit evaluations of the action of the homotopy operator on the various forms are clearly essential. Since the choice of the point $\{X_o^a\}$ does not result in essential changes in the theory, as shown by the form invariance of the resulting relations under changes in that choice, we select $X_o^a = 0$ in the interests of simplicity. In this case, (6.6) becomes

(7.1) $$\mathcal{X} = X^A \partial_A + T\partial_4 \ .$$

It is then simply a matter of applying the definition (6.7) to the various forms involved.

We start with the form of highest degree, namely

(7.2) $$\Omega^i = - S^i \wedge dT + \Theta^i = - S^{Ai}\ \mu_A \wedge dT + \theta^i \mu \ .$$

Since Ω^i is a form of degree 3, (6.7) and (4.4) yield

(7.3) $$H\langle \Omega^i \rangle = \int_o^1 \lambda^2 \{ -X^B \tilde{S}^{Ai} \mu_{BA} \wedge dT + (X^A \tilde{\theta}^i - T\tilde{S}^{Ai})\mu_A \}\ d\lambda \ ,$$

where

$$\mu_{BA} = \partial_B \rfloor \mu_A = \frac{1}{2}\ \partial_B \rfloor (\bar{e}_{AFG} dX^F \wedge dX^G) = - e_{BAC} dX^C \ .$$

Thus, if we define the linear integral operator h_2 by

(7.4) $$h_2\langle \rho \rangle (X^a) = \int_o^1 \lambda^2\ \rho(\lambda X^a)\ d\lambda \ ,$$

we have

$$H\langle \Omega^i \rangle = X^B\, h_2\langle S^{Ai}\rangle e_{BAC}\; dX^C \wedge dT + (X^A\, h_2\langle \theta^i\rangle + T\, h_2\langle S^{Ai}\rangle)\, \mu_A \; . \tag{7.5}$$

For forms of degree 2, a similar calculation and (4.5), (4.14) show that

$$H\langle \mathcal{D}^i \rangle = X^A\, h_1\langle J^i_A\rangle\, dT - (T\, h_1\langle J^i_C\rangle + X^B\, h_1\langle \alpha^{Ai}\rangle e_{ABC})\; dX^C \; , \tag{7.6}$$

$$H\langle \mathcal{K}^i \rangle = -\, X^A\, h_1\langle \omega^i_A\rangle\, dT - (-T\, h_1\langle \omega^i_C\rangle + X^B\, h_1\langle k^{Ai}\rangle e_{BAC})\; dX^C \; , \tag{7.7}$$

where h_1 is the linear integral operator

$$h_1\langle \rho\rangle(X^A) = \int_0^1 \lambda\; \rho(\lambda X^a)\; d\lambda \; . \tag{7.8}$$

For forms of degree 1, (4.13) yields

$$H\langle \mathcal{B}^i \rangle = T\, h\langle V^i\rangle + X^A\, h\langle \mathcal{B}^i_A\rangle \tag{7.9}$$

with

$$h\langle \rho\rangle(X^a) = \int_0^1 \rho(\lambda X^a)\; d\lambda \; . \tag{7.10}$$

These results and (6.10), (6.15), (6.16) give the evaluations

$$\bar{\mathcal{K}}^i = X^B\, h_2\langle S^{Ai}\rangle E_{BAC}\, dX^C \wedge dT + (X^A\, h_2\langle \theta^i\rangle + T\, h_2\langle S^{Ai}\rangle)\mu_A \; , \tag{7.11}$$

$$(7.12)\qquad \mathcal{B}^i = X^A\, h_1\langle J^i_A\rangle dT - (T\, h_1\langle J^i_C\rangle + X^B\, h_1\langle \alpha^{Ai}\rangle E_{BAC})\, dX^C .$$

$$(7.13)\qquad \chi^i = T\, h\langle V^i\rangle + X^A\, h\langle \beta^i_A\rangle .$$

Since (6.14) may be written in the equivalent forms

$$(7.14)\qquad K^i = \bar{K}^i + dH\langle K^i\rangle = -\,\omega^i \wedge dT + k^i$$

$$(7.15)\qquad \mathcal{B}^i = d\chi^i + H\langle \mathcal{D}^i - K^i\rangle = V^i dT + \beta^i$$

we obtain explicit expressions for the forms ω^i and k^i , while (7.15) yields the relations

$$(7.16)\qquad V^i = \partial_4 \chi^i + X^A\, h_1\langle J^i_A + \omega^i_A\rangle ,$$

$$(7.17)\qquad \beta^i_C = \partial_C \chi^i - T\, h_1\langle J^i_C + \omega^i_C\rangle - X^B\, h_1\langle \alpha^{Ai} - k^{Ai}\rangle e_{BAC} .$$

In the absence of defects, J^i_A , ω^i_A , α^{Ai} and k^{Ai} all vanish and (7.16), (7.17) give

$$(7.18)\qquad \overset{T}{V}{}^i = \partial_4 \chi^i \ , \quad \overset{T}{\beta}{}^i_C = \partial_C \chi^i ,$$

where the superscript T designates the total that results from the integrable displacement field $\chi^i = \chi^i(X^a)$. We may thus write

$$(7.19)\qquad \overset{T}{\mathcal{B}}{}^i = \overset{T}{V}{}^i\, dT + \overset{T}{\beta}{}^i = d\chi^i .$$

When defects are present, it is customary to write [10, 17]

(7.20) $\overset{T}{B}{}^i = B^i + \overset{P}{B}{}^i$

with

(7.21) $\overset{P}{B}{}^i = \overset{P}{V}{}^i \, dT + \overset{P}{\beta}{}^i$

where $\overset{P}{V}{}^i$ are the components of the *plastic velocity* and $\overset{P}{\beta}{}^i = \overset{P}{\beta}{}^i_C \, dX^C$ as the *plastic distortion* 1-forms. Now, it is clear from $\beta^i = d\chi^i + H\langle \mathcal{D}^i - K^i \rangle$ that the part $H\langle \mathcal{D}^i - K^i \rangle$ is determined by the defects while the integrable part $d\chi^i$ is not. Indeed, if $d\chi^i$ were to be determined by the defects, there would be no undetermined variables that could be used to secure satisfaction of the laws of balance of momentum and energy. It is thus necessary that the terms $d\chi^i$ remain undetermined by the defects, in which case we may consider that $x^i = \chi^i(X^a)$ continues to define the total deformation of the body. This, however, is the case only if we identify $d\chi^i$ with $\overset{T}{B}{}^i$ when defects are present. When $d\chi^i = \overset{T}{B}{}^i$ is substituted into (7.20), we obtain

(7.22) $d\chi^i = B^i + \overset{P}{B}{}^i$

and comparison with (7.15) gives the explicit evaluation

(7.23) $\overset{P}{B}{}^i = - H\langle \mathcal{D}^i - K^i \rangle$.

Thus, (7.16), (7.17) and (7.21) give the specific evaluations

(7.24) $\overset{P}{V}{}^i = - X^A \, h_1\langle \omega^i_a + J^i_A \rangle$,

$$(7.25)\qquad {}^{P}\beta^{i}_{C} = T\, h_1\langle J^i_A + \omega^i_A\rangle + X^B\, h_1\langle \alpha^{Ai} - k^{Ai}\rangle e_{BAC} .$$

We have thus shown that *the plastic distortion is uniquely determined by the dislocation 2-forms* $\mathcal{D}^i$ *and the spin-twist 2-forms* $\mathcal{K}^i$.

A very important conclusion now emerges. It follows directly from (7.23) that there is no plastic distortion whenever $H\langle \mathcal{D}^i - \mathcal{K}^i\rangle = 0$; that is, whenever $\mathcal{D}^i - \mathcal{K}^i$ belongs to the kernel of the linear operator H . However, $dB^i = \mathcal{D}^i - \mathcal{K}^i$, so that $\mathcal{D}^i - \mathcal{K}^i$ is a system of exact 2-forms while $H\langle H\langle\rho\rangle\rangle \equiv 0$, $\rho \equiv dH\langle\rho\rangle + H\langle d\rho\rangle$ show that exact forms and forms in the kernel of H intersect only in the zero forms (see [3] Corollary 5-6.3). Thus, $H\langle \mathcal{D}^i - \mathcal{K}^i\rangle = 0$ only if $\mathcal{D}^i = \mathcal{K}^i$. It is convenient to refer to defects with these properties as *selfequilibrating* since they give rise to identically vanishing plastic velocities and distortions.

There exist nontrivial selfequilibrating defects that possess identically vanishing plastic velocities and distortions. The set of all such defects are those that satisfy the relations

$$(7.26)\qquad \mathcal{D}^i = \mathcal{K}^i ;$$

that is

$$(7.27)\qquad J^i_A = -\,\omega^i_A , \quad \alpha^{Ai} = k^{Ai} .$$

We pause at this point in order to establish a result that leads to certain simplifications. Equation (7.8) shows that

$$X^b\, h_1\langle\rho_b\rangle = \int_0^1 X^b \lambda\ \rho_b(\lambda X^a)\, d\lambda = \int_0^1 (X^b \rho_b)(\lambda X^a)\, d\lambda\ ,$$

and hence (7.10) gives

$$(7.28)\qquad X^b\, h_1\langle\rho_b\rangle = h\langle X^b \rho_b\rangle\ .$$

When this is used in conjunction with (7.24) and (7.25), we obtain

$$(7.29)\qquad \overset{P}{V}{}^i = -\, h\langle X^A (J^i_A + \omega^i_A)\rangle\ ,$$

$$(7.30)\qquad \overset{P}{\beta}{}^i_C = h\langle T(J^i_C + \omega^i_C) + X^B(\alpha^{Ai} - k^{Ai})e_{BAC}\rangle\ .$$

We may thus draw the following conclusions: If the defects are such that

$$(7.31)\qquad X^A(J^i_A + \omega^i_A) = 0$$

then there is no plastic velocity. If the defects are such that

$$(7.32)\qquad T(J^i_A + \omega^i_A) = X^B(k^{Ai} - \alpha^{Ai})e_{BAC}$$

then there is no plastic distortion. Sufficient conditions for the presence of at least one nonzero component of $\overset{P}{B}{}^i$ *when there are defects presence are* $\mathcal{D}^i \not\equiv 0\ ,\ K^i \equiv 0$ *or* $K^i \not\equiv 0\ ,\ \mathcal{D}^i \equiv 0$. These conclusions would seem to have serious implications in interpretations of plastic responses of materials in terms of defect mechanisms.

There is one further aspect of the homotopy operator, H , that must be noted. This operator gives rise to the operators $h_2\langle\,\rangle$, $h_1\langle\,\rangle$ and $h\langle\,\rangle$ where

$$(7.33) \qquad h_k\langle\rho\rangle(X^A,T) = \int_0^1 \lambda^k \, \rho(\lambda X^A, \lambda T)\, d\lambda \; .$$

Thus, the evaluation of $h_k\langle\rho\rangle$ at any space-time point $\{X^a\} = \{X^A,T\}$ with $T>0$ involves a weighted integration over the line from $\{X_o^a\} = \{0^a\}$ to $\{X^A,T\}$; that is, it involves the values of ρ for all values of the time between $X^4 = 0$ and $X^4 = T$. The theory thus exhibits both spatial and temporal non-locality!

8. The Analogy with Electrodynamics

It has been pointed out that the gauge transformation properties of the field equations of defects were first discovered by A. A. Golebiewska-Lasota through an argument by analogy between the field equations of electrodynamics and the field equations of dislocations [33]. The 4-dimensional formulation of defect dynamics presented above allows us to make an exact analogy between defect dynamics and electrodynamics, as we now proceed to show. There are, however, essential differences between the two theories, for the analogy requires more than one electromagnetic field and the presence of magnetic charges and currents if disclinations are present, while the respective constitutive theories are entirely different. The presentation of electrodynamics with magnetic charges and currents will be taken directly from [42], to which the reader is referred for the details.

Let

$$(8.1) \qquad J = J^A \mu_A \wedge dT - q \mu$$

denote the 3-form of *electric charge-current* and let

$$(8.2) \qquad G = G^A \mu_A \wedge dT - g \mu$$

denote the 3-pseudoform of *magnetic charge-current*. The laws of conservation of electric charge and of magnetic charge are given by the exterior equations

$$(8.3) \qquad dJ = 0 \quad , \quad dG = 0 \ ,$$

respectively. These exterior equations may be integrated directly and we obtain

$$(8.4) \qquad J = dH \quad , \quad G = dF$$

where H is the 2-form of electric charge-current potential and F is the 2-pseudoform of magnetic charge-current potential. If we introduce the standard electromagnetic field vectors $\vec{E}$, $\vec{H}$, $\vec{B}$, $\vec{D}$ through the relations

$$(8.5) \qquad H = H_A \ dX^A \wedge dT - D^A \ \mu_A \ ,$$

$$(8.6) \qquad F = - E_A \ dX^A \wedge dT - B^A \ \mu_A \ ,$$

then (3.11), (3.12), (8.1), (8.2) show that (8.4) are the same as Maxwell's equations

$$(8.7) \qquad \vec{\nabla}\times\vec{H} - \partial_4\vec{D} = \vec{J} \quad , \quad \vec{\nabla}\cdot\vec{D} = q \ ,$$

$$(8.8) \qquad - \vec{\nabla}\times\vec{E} - \partial_4\vec{B} = \vec{G} \quad , \quad \vec{\nabla}\cdot\vec{B} = g$$

in the presence of both electric and magnetic charges. A further integration of the second of (8.4) yields

$$(8.9) \qquad F = - dA + H\langle G\rangle \ .$$

Let us now simply observe that $dJ = 0$, $dG = 0$ are of exactly the same form as $dP_i = 0$, $dP_4 = 0$ and $d\Omega^i = 0$, only that there are more of the latter.

There is one essential difference between the systems $dP_i = 0$, $dP_4 = 0$ on the one hand and the system $d\Omega^i = 0$ on the other; for we always must balance linear momentum and energy, while we need balance disclinations only if there are defects present that represent disclinations. Further J is a 3-form and G is a 3-pseudoform, while momentum and energy possess vector and scalar transformation properties while disclinations are associated with symmetry breaking of the rotation group and thus are characterized by axial or pseudo quantities. With these observations in mind, it seems reasonable to identify the four 3-forms P_i , P_4 with four distinct 3-forms of electric charge-current J_i , J_4 and to identify the three 3-forms Ω^i with three distinct 3-pseudoforms of magnetic charge current G^i . This gives the exact analogies

$$\vec{\nabla}\times\vec{H}_i - \partial_4\vec{D}_i = \vec{\sigma}_i \quad , \quad \vec{\nabla}\cdot\vec{D}_i = -\, p_i \tag{8.10}$$

$$\vec{\nabla}\times\vec{H}_4 - \partial_4\vec{D}_4 = \vec{W} \quad , \quad \vec{\nabla}\cdot\vec{D}_4 = -\, e \tag{8.11}$$

for momentum and energy and

$$-\,\vec{\nabla}\times\vec{E} - \partial_4\vec{B} = \vec{S} \quad , \quad \vec{\nabla}\cdot\vec{B} = -\,\Theta \tag{8.12}$$

for disclinations; that is

$$P_i = J_i \quad , \quad P_4 = J_4 \quad , \quad \Omega^i = G^i \quad . \tag{8.13}$$

As a result of this exact analogy, we obtain the following conclusion. *There is an exact analogy between classical electrodynamics* (electrodynamics without magnetic currents) *and defect dynamics only when the disclination densities and currents vanish identically.* A nonvanishing disclination density or current requires a corresponding nonvanishing magnetic charge or current.

There is a useful further analogy, namely that

(8.14) $\quad \mathcal{D}^{i} = F^{i}$

that follows from $d\mathcal{D}^{i} = \Omega^{i} = G^{i} = dF^{i}$. When (4.5) and (8.6) are used, (8.14) gives the following explicit analogies

(8.15) $\quad J_{A}^{i} = - E_{A}^{i} \ , \quad \alpha^{Ai} = - B^{Ai} \ .$

It is then a straightforward matter to use (8.12) and (8.15) to obtain the standard conclusions that dislocations may only terminate on disclinations, etc.

The emergence of a 45-fold Abelian gauge group for defect dynamics, as opposed to a 4-fold gauge group for electrodynamics now seems reasonable by analogy, for defect dynamics requires three co-occupying $\check{\dot{\vec{E}}}$ and $\vec{B}$ and four co-occupying $\vec{H}$ and $\vec{D}$ fields, and electrodynamics usually deals with given J and G forms while defect dynamics requires that the corresponding

P's and Ω's be obtained by means of constitutive relations.

A strong note of caution must be sounded in this vein, for one can not solve the equations of defect dynamics by solving simultaneous systems of classical electromagnetic field equations. This is easily seen by a comparison of the constitutive theory. In electrodynamics, the constitutive relations tell us that $\vec{D} = \varepsilon_o(\vec{E} + \vec{\text{Polarization}})$, $\vec{H} = \mu_o^{-1}(\vec{B} + \vec{\text{Magnetization}})$, while even in the absence of defects, the $\vec{\sigma}_i$'s are functions of the strain measures that are formed from $\partial_A \chi^i$ that come from the β's . Thus, since the $\vec{\sigma}_i$'s are identified with electric current vectors while the β's correspond to the vector potentials, an exact analogy would require the electromagnetic fields to yield electric currents that are determined by the vector potentials of the fields; a situation that certainly does not obtain with classical electrodynamics.

9. Thermodynamic Considerations and the Resulting Constitutive Equations

The field equations given in Section 2 serve to determine the time derivatives of the quantities p_i , α^{Ai} , θ^i , k^{Ai} , β^i_A and e through the relations

(9.1) $$\partial_4 p_i = \partial_A \sigma^A_i \ , \quad \partial_4 \sigma^{Ai} = -e^{ABC} \partial_B J^i_C - S^{Ai} \ ,$$

(9.2) $$\partial_4 \theta^i = - \partial_A S^{Ai} \ , \quad \partial_4 k^{Ai} = e^{ABC} \partial_B \omega^i_C - S^{Ai} \ ,$$

(9.3) $$\partial_4 \beta^i_A = \partial_A V^i - J^i_A - \omega^i_A \ ,$$

and

(9.4) $$\partial_4 e = \partial_A W^A \ .$$

We concentrate on the latter equation for the thermodynamic considerations.

Since e is the density of total energy (the Hamiltonian density) it may be considered as the sum of the kinetic energy density (expressed in terms of the momenta p_i) and the potential or internal energy of the dynamic process at hand. We use the concept of a companion constrained equilibrium thermostatic state of Bataille and Kestin [43] in order to obtain a representation of the contribution from the internal energy: *Any instantaneous nonequilibrium state can be approximated with sufficient accuracy by a constrained equilibrium state with the same values of the instantaneous mechanical substate variables to allow use of the temperature,* T *,* *the specific entropy,* s *, and the internal energy* $U(s;\ldots)$ *as characteristic of the instantaneous non-equilibrium state.* This motivates the constitutive assumption

$$(9.5) \qquad e = e(s, p_i, \beta^i_A, \alpha^{Ai}, k^{Ai}, \theta^i) \ .$$

The *thermodynamic temperature* is then defined by the standard thermostatic statement

$$(9.6) \qquad T = \partial e/\partial s$$

while the components of *kinematic velocity* are defined by the standard Hamiltonian relations

$$(9.7) \qquad v^i = \partial e/\partial p_i \ .$$

It is now necessary to give constitutive assumptions concerning the evaluation of the quantities W^A in order that (9.4) may be used as a basis for the practices of nonequilibrium thermodynamics. Let

$$(9.8) \qquad Q = Q^A \mu_A$$

denote the 2-form of nonmechanical energy (heat) influx. Since v^i are the components of the kinematic velocity, $v^i \, \sigma^A_i \mu_A$ is the 2-form of the rate of stress-work. It thus seems reasonable to make the constitutive assumption

(9.9) $$W^A = - Q^A + v^i \sigma_i^A ;$$

that is, the 2-form W has the evaluation

(9.10) $$W = - Q + v^i \sigma_i .$$

It should be realized, however, that (9.9) explicitly assumes that there are no couple stresses on boundaries and that there is no rate of tork-work associated with such couple stresses. There is indeed a serious question involved here if the material admits nontrivial nonselfequilibrating disclinations, that seems best left to a future paper. At any rate, we have given a well defined set of constitutive assumptions and will proceed with these for the remainder of the paper.

What now remains is an implementation of the second half of the second law of thermodynamics; namely, that the rate of entropy production be nonnegative for all processes of the body. When (9.5) through (9.9) are substituted into (9.4) and (9.1) through (9.3) are used to eliminate the time derivatives, a straightforward sequence of manipulations leads to the relation

(9.11) $$\partial_4 s + \partial_B \eta^B = \dot{\Theta} .$$

Here

$$(9.12)\qquad \eta^{B} = \frac{1}{T}\left\{Q^{B} - e^{ABC}\left(\frac{\partial e}{\partial \alpha^{Ai}} J^{i}_{C} - \frac{\partial e}{\partial k^{Ai}} \omega^{i}_{C}\right) - \frac{\partial e}{\partial \theta^{i}} S^{Bi}\right\}$$

are the components of the entropy flux vector and

$$(9.13)\qquad \dot{\Theta} = -\frac{1}{T}\eta^{B}\partial_{B}T + \frac{1}{T}\left(\sigma^{A}_{i}\partial_{A}v^{i} - \frac{\partial e}{\partial \beta^{i}_{A}}\partial_{A}V^{i}\right)$$

$$+ \frac{1}{T}\left\{\frac{\partial e}{\partial \beta^{i}_{C}} - e^{ABC}\partial_{B}\left(\frac{\partial e}{\partial \alpha^{Ai}}\right)\right\}J^{i}_{C}$$

$$+ \frac{1}{T}\left\{\frac{\partial e}{\partial \beta^{i}_{C}} + e^{ABC}\partial_{B}\left(\frac{\partial e}{\partial k^{Ai}}\right)\right\}\omega^{i}_{C}$$

$$+ \frac{1}{T}\left\{\frac{\partial e}{\partial \alpha^{Ai}} + \frac{\partial e}{\partial k^{Ai}} - \partial_{A}\left(\frac{\partial e}{\partial \theta^{i}}\right)\right\}S^{Ai}$$

is the entropy production. The second half of the second law requires that

$$(9.14)\qquad \dot{\Theta} \geqq 0$$

for all possible processes within the body.

It is clear from the start that we have two velocity fields to contend with; namely the kinematic field $\{v^{i}\}$ and the field $\{V^{i}\}$. This is reminiscent of the fact that the integrable velocity field $\{\overset{T}{V}{}^{i}\}$ is given as the sum $\{V^{i} + \overset{P}{V}{}^{i}\}$ of the $\{V^{i}\}$ field and the plastic velocity field $\{\overset{P}{V}{}^{i}\}$. In fact, since $\{\overset{T}{V}{}^{i}\}$ is the total velocity field of the body, Newtonian mechanics would require that we set

$$v^i = \overset{T}{V}{}^i = V^i + \overset{P}{V}{}^i , \tag{9.15}$$

in which case we would have

$$p^i = \rho_o \delta_{ij}(V^j + \overset{P}{V}{}^j) \tag{9.16}$$

as the total momentum that enters into the balance of linear momentum equations. In order to cover the various possible contingencies, let us put

$$v^i = \partial e/\partial p_i = V^i + \gamma^i . \tag{9.17}$$

In this instance, (9.13) becomes

$$\begin{aligned} T\dot{\Theta} = & - \eta^B \partial_B T + \left(\sigma_i^A - \frac{\partial e}{\partial \beta_A^i}\right)\partial_A V^i + \sigma_i^A \partial_A \gamma^i \\ & + \left\{\frac{\partial e}{\partial \beta_C^i} - e^{ABC}\partial_B\left(\frac{\partial e}{\partial \alpha^{Ai}}\right)\right\} J_C^i \\ & + \left\{\frac{\partial e}{\partial \beta_C^i} + e^{ABC}\partial_B\left(\frac{\partial e}{\partial k^{AC}}\right)\right\} \omega_C^i \\ & + \left\{\frac{\partial e}{\partial \alpha^{Ai}} + \frac{\partial e}{\partial k^{Ai}} - \partial_A\left(\frac{\partial e}{\partial \theta^i}\right)\right\} S^{Ai} . \end{aligned} \tag{9.18}$$

This is the general form of the entropy production that has to be rendered nonnegative by the constitutive equations of the theory.

The usual practices of defect dynamics assume that the stress arises only as a consequence of the variables β^{i}_{C} , not through the total distortion $\overset{T}{\beta}{}^{i}_{C} = \beta^{i}_{C} + \overset{P}{\beta}{}^{i}_{C}$; that is, the plastic distortion does not give rise to resulting tractions (i.e., is "stress free"). It is thus customary to assume that

$$(9.19) \qquad \sigma^{A}_{i} = \partial e / \partial \beta^{i}_{A} \;;$$

a direct generalization of the existence of a strain energy function in classical elasticity theory. When this is substituted into (9.18), we have

$$(9.20) \qquad T\dot{\Theta} = -\,\eta^{B}\partial_{B}T + \sigma^{A}_{i}\partial_{A}\gamma^{i}$$
$$+ \left\{\sigma^{C}_{i} - e^{ABC}\partial_{B}\left(\frac{\partial e}{\partial \alpha^{Ai}}\right)\right\} J^{i}_{C}$$
$$+ \left\{\sigma^{C}_{i} + e^{ABC}\partial_{B}\,\frac{\partial e}{\partial k^{Ai}}\right\} \omega^{i}_{C}$$
$$+ \left\{\frac{\partial e}{\partial \alpha^{Ai}} + \frac{\partial e}{\partial k^{Ai}} - \partial_{A}\left(\frac{\partial e}{\partial \Theta^{i}}\right)\right\} S^{Ai}$$

There is one aspect of the theory that remains outstanding, for we have as yet said nothing about the total displacement $\chi^{i}(X^{A},T)$. However, we do have the relations

$$v^{i} = V^{i} + \gamma^{i} \quad \text{and} \quad \partial_{4}\chi^{i} = V^{i} + \overset{P}{V}{}^{i} \;,$$

and hence an elimination of the common terms V^{i} yield

(9.21) $\quad v^i - \partial_4 \chi^i = \gamma^i - \overset{P}{V}{}^i$.

On the other hand, the equations of defect dynamics do not serve to determine the functions χ^i , so we may use the primitive definition of total velocity in Newtonian mechanics as the time derivative of the total displacement function to set

(9.22) $\quad v^i = \partial_4 \chi^i$.

In this event, the equations of balance of linear momentum will serve to determine the quantities χ^i and (9.21) gives

(9.23) $\quad \gamma^i = \overset{P}{V}{}^i$.

When these results are substituted into (9.20), we obtain

(9.24)
$$T\dot{\theta} = - \eta^B \partial_B T + \sigma_i^A \partial_A \overset{P}{V}{}^i + \left\{\sigma_i^C - e^{ABC} \partial_B \left(\frac{\partial e}{\partial \alpha^{Ai}}\right)\right\} J_C^i + \left\{\sigma_i^C + e^{ABC} \partial_B \left(\frac{\partial e}{\partial k^{Ai}}\right)\right\} \omega_C^i + \left\{\frac{\partial e}{\partial \alpha^{Ai}} + \frac{\partial e}{\partial k^{Ai}} - \partial_A \left(\frac{\partial e}{\partial \theta^i}\right)\right\} S^{Ai} .$$

It is now a simple matter to identify the thermodynamic fluxes and forces through the right-hand side of (9.24)

in order to use the standard practices of nonequilibrium thermodynamics [43,44] to obtain a complete system of constitutive relations. The results contained in (9.24) are to some extent remarkable, for the occurrence of the terms $\sigma_i^A \partial_A \overset{P}{V}{}^i$ provide a simple and direct means of implementing the notion of a yield surface even though this concept has not been used anywhere in the theory up to this point.

10. Examples Modeling the Plastic State

A specific example is useful at this point in order to see just what is possible with a "bare bones" application of the theory. We consider the case in which

(10.1) $\sigma_i^A = \partial e/\partial \beta_i^A \quad , \quad v^i = \partial_4 \chi^i$

and

(10.2) $e = e(s, p_i, \beta_A^i) = \frac{1}{2\rho_o} p_i \delta^{ij} p_j + \rho_o U(s, \beta_A^i)$.

Thus, $\partial e/\partial \alpha^{Ai} = 0$, $\partial e/\partial k^{Ai} = 0$, $\partial e/\partial \theta^i = 0$ and (9.7) together with (10.1) give $p_i = \rho_o \delta_{ij} \partial_4 \chi^j$. If we also assume that the body is free of disclination effects, so that we may put $S^{Ai} = 0$, $\omega_A^i = 0$, $\theta^i = 0$, then (9.12) and (9.24) give

(10.3) $T\eta^A = Q^A$,

(10.4) $T\dot{\Theta} = -\eta^B \partial_B T + \sigma_i^A \partial_A \overset{P}{V}{}^i + \sigma_i^C J_C^i$.

If we choose

(10.5) $Q^B = -K^{BC} \partial_C T$,

then (10.3) shows that

$$-\eta^B \partial_B T = \frac{1}{T} K^{BC} \partial_B T \partial_C T$$

is positive provided $((K^{BC}))$ is a positive definite matrix, in which case (10.5) is just the Fourier law of heat conduction. In the interests of simplicity, we shall neglect thermal conduction effects since most theories of the plastic state do exactly that. This amounts to putting $K^{BC} = 0$, in which case (10.4) becomes

$$T\dot{\Theta} = \sigma_i^A \, \partial_A \overset{P}{V}{}^i + \sigma_i^C \, J_C^i \, . \tag{10.6}$$

Now, $\partial_A \chi^i = \beta_A^i + \overset{P}{\beta}{}_A^i$, $\partial_4 \chi^i = V^i + \overset{P}{V}{}^i$ yield $\beta_A^i = \partial_A \chi^i - \overset{P}{\beta}{}_A^i$, $V^i = \partial_4 \chi^i - \overset{P}{V}{}^i$. Thus, since $\omega_A^i = 0$ here, (9.3) yields the relations

$$\partial_4 \overset{P}{\beta}{}_A^i = J_A^i + \partial_A \overset{P}{V}{}^i \tag{10.7}$$

and the rate at which work is done by the plastic distortion is given by

$$\sigma_i^A \, \partial_4 \overset{P}{\beta}{}_A^i = \sigma_A^i (J_A^i + \partial_A \overset{P}{V}{}^i) = T\dot{\Theta} \geq 0 \, . \tag{10.8}$$

Drucker's postulate in differential form [36] is equivalent to the statement that the rate at which work is done by the plastic distortion is nonnegative. *The models considered here are those for which Drucker's postulate is satisfied whenever the second half of the second law of thermodynamics is satisfied* $(\dot{\Theta} \geq 0)$.

It is well known [36] that Drucker's postulate implies the existence and convexity of a loading surface and an associated plastic flow law. Similar results may also be derived directly from the practices of non-equilibrium thermodynamics. The basic problem is that of securing the non-negativity of the form

$$(10.9) \qquad T\dot{\Theta} = \sigma_i^A (J_A^i + \partial_A \overset{P}{V}{}^i) \ .$$

Once e is given, $\sigma_i^A = \partial e / \partial \beta_A^i$ is known, so we identify σ_i^A with the thermodynamic forces. The corresponding thermodynamic fluxes are thus identified by (10.9) to be $J_A^i + \partial_A \overset{P}{V}{}^i = \partial_4 \overset{P}{\beta}{}_A^i$. The results given in Theorem B of [45] show that $J_A^i + \partial_A V^i$ will be such as to render $\dot{\Theta} \geq 0$ if and only if there exist a nonnegative function $u(\sigma_i^A;T,\ldots)$, a scalar valued function $\phi(\sigma_i^A;T,\ldots)$ such that

$$(10.10) \qquad \sigma_i^A \, \partial\phi/\partial\sigma_i^A \geq 0 \ ,$$

and a collection of functions U_A^i such that

$$(10.11) \qquad \sigma_i^A \, U_A^i = 0 \ ,$$

in which case

$$(10.12) \qquad J_A^i + \partial_A \overset{P}{V}{}^i = u\{\partial\phi/\partial\sigma_i^A + U_A^i\} \ .$$

When this is substituted back into (9.33), we obtain

$$u \, \sigma_i^A \, \partial\phi/\partial\sigma_i^A = T\dot{\Theta}$$

and hence

$$(10.13) \qquad u = \frac{T\dot{\Theta}}{\sigma^B_j \, \partial\phi/\partial\sigma^B_j} \; .$$

Thus, (10.12) yields

$$(10.14) \qquad J^i_A + \partial_A \overset{P}{V}{}^i = \frac{T\dot{\Theta}}{\sigma^B_j \, \partial\phi/\partial\sigma^B_j} \{\partial\phi/\partial\sigma^A_i + U^i_A\}$$

and hence (10.7) becomes the plastic flow law

$$(10.15) \qquad \partial_4 \overset{P}{\beta}{}^i_A = \frac{T\dot{\Theta}}{\sigma^B_j \, \partial\phi/\partial\sigma^B_j} \{\partial\phi/\partial\sigma^A_i + U^i_A\}$$

If we assume that the nondissipative part, U^i_A, of $J^i_A + \partial_A \overset{P}{V}{}^i$ vanishes identically, as is the case with most analyses given in the literature, then (10.15) assumes the familiar form

$$(10.16) \qquad \partial_4 \overset{P}{\beta}{}^i_A = \frac{T\dot{\Theta}}{\sigma^B_j \, \partial\phi/\partial\sigma^B_j} \partial\phi/\partial\sigma^A_i = J^i_A + \partial_A \overset{P}{V}{}^i \; .$$

In this case $\phi(\sigma^A_i;T,\ldots) =$ constant defines the instantaneous loading surface, the existence of which is implied by satisfaction of the second half of the second law of thermodynamics.

It is of importance to note that the occurrence of $\dot{\Theta}$ in (10.14) through (10.16) provide for both loading and unloading situations, for $\dot{\Theta}$ would vanish identically in unloading if the unloading is assumed to occur reversibly. Further, since (10.10) shows that ϕ would reduce in value during an incrimental unloading process, the unloading process could be modeled by choosing

$$(10.17)\qquad \dot{\Theta} = \begin{cases} F(\phi;\ ,\ldots) & \text{for} \quad \partial_4\phi \geq 0\ , \\ 0 & \text{for} \quad \partial_4\phi < 0\ , \end{cases}$$

with F positive valued and monotone increasing in ϕ for all values of the other arguments.

There is one aspect of the model from the dislocation point of view that is still unresolved, for (10.14) only serves to determine the sum $J^i_A + \partial_A \overset{P}{V}{}^i$, not J^i_A and $\partial_A \overset{P}{V}{}^i$ separately. One obvious resolution is to take $\partial_A \overset{P}{V}{}^i$ to be the exact part of the right-hand side of (9.37) when considered as coefficients of a 1-form. In this event, J^i_A would then be identified with the antiexact part.

Another possibility is to take

$$(10.18)\qquad \partial_A \overset{P}{V}{}^i = \frac{T\dot{\Theta}}{\sigma^B_j\ \partial\phi/\partial\sigma^B_j}\, U^i_A$$

where $\dot{\Theta}$, ϕ and U^i_A are now to be chosen so as to secure satisfaction of (10.11) and conditions that are implied by the fact that the left-hand side of (10.18) is a gradient with respect to the X's. This has the pleasant circumstance of securing the condition

$$(10.19)\qquad \sigma^A_i\ \partial_A \overset{P}{V}{}^i \equiv 0\ ,$$

as follows from (10.11), in which case the dislocation current is given by

$$(10.20) \qquad J^i_A = \frac{T\dot{\Theta}}{\sigma^B_j \; \partial\phi/\partial\sigma^B_j} \, \partial\phi/\partial\sigma^A_i \; .$$

Under these circumstances, *the gradient of the plastic velocity is nondissipative* in the sense that $\sigma^A_i \, \partial_A \overset{P}{V}{}^i$ makes an identically zero contribution to the entropy production in all possible processes. The quantity $\phi(\sigma^A_i; T, \ldots)$ may now be interpreted as a *dislocation current potential,* and (9.43) gives a direct correlation between the state of stress and the entropy production with the dislocation current; *the dislocation current is orthogonal to the surface* ϕ = *constant in stress space.* Again, since $\dot{\Theta} = 0$ in an elastic unloading process, the dislocation current will vanish in an elastic unloading process.

The above remarks, although encouraging, are not definitive. In fact, it would seem that an answer to the question of the disentanglement of J^i_A and $\partial_A \overset{P}{V}{}^i$ will be dependent upon exactly what phenomena are to be modeled by the theory, and thus should await a detailed treatment of specific problems associated with materials with definite response properties.

References

1. Ślebodziński, W., *Exterior Forms and Their Applications* (Polish Scientific Publishers, Warsaw, 1970).

2. Cartan, E., *Les systemes differentials exterieurs et leurs applications geometriques* (Hermann, Paris, 1945).

3. Edelen, D. G. B., *Isovector Methods For Equations of Balance* (Sijthoff & Noordhoff, Alphen aan den Rijn, The Netherlands, 1980).

4. Sternberg, S., *Lectures on Differential Geometry* (Prentice-Hall, Englewood Cliffs, 1964).

5. Edelen, D. G. B., Int. J. Engng. Sci. 18 (1980), 1095.

6. Yang, C. N., and R. L. Mills, Phys. Rev. 96 (1954), 191.

7. Actor, A., Rev. Mod. Phys. 51 (1979), 461.

8. Drechsler, W. and M. E. Mayer, *Fiber Bundle Techniques in Gauge Theories* (Lecture Notes in Physics No. 67, Springer, Berlin, 1977).

9. Yang, C. N., *Proceedings of the Sixth Hawaii Topical Conference in Particle Physics*, ed. by P. N. Dodson, Jr., University of Hawaii and Manoa, Honolulu, (1975).

10. Kröner, E., "Continuum Theory of Defects," Series of Lectures held at the Summer School on the Physics of Defects, Les Houches 1980, to appear.

11. Peach, M. and J. S. Koehler, Phys. Rev. 80 (1950), 436.

12. Edelen, D. G. B., Int. J. Engng. Sci. 17 (1979), 441.

13. Edelen, D. G. B., Ann. Phys. (N.Y.) 133 (1981), 286.

14. Rogula, D., *Trends in Applications of Pure Mathematics to Mechanics*, ed. by G. Fishera (Pitman, London, 1976), 311.

15. Eshelby, J. D., J. Elasticity 5 (1975), 321.

16. Yang, C. N., and T. T. Wu, *Properties of Matter Under Unusual Conditions*, ed. H. Mark and S. Fernbach (Interscience Publishers, New York, 1969).

17. Kossecka, E., Arch. Mech. 27 (1975), 79.

18. Zorawski, M., *Theorie Mathematique des Dislocation* (Dunod, Paris, 1967).

19. Kröner, E., "Dislocations and Continuum Mechanics (original published in the Aug. 1962 issue of AMR).

20. Nabarro, F. R. N., *Theory of Crystal Dislocations* (Oxford University Press, Oxford, 1967).

21. Weertman, J. and J. R. Weertman, *Elementary Dislocation Theory* (The Macmillan Company, New York, 1964).

22. Volterra, V., Ann. Ecole Norm. Super. 24 (1907), 401.

23. Weingarten, G., Atti. Accad. naz. Lincii, Rend., Cl. Sci. fis. mat. natur. 5 (1901), 57.

24. Somigliana, C., Atti. Accad. naz. Lincii, Rend., Cl. Sci. fis. mat. natur. 23 (1914), 463.

25. Orowan, E., Z. Phys. 89 (1934), 605.

26. Polanyi, M., Z. Phys. 89 (1934), 660.

27. Taylor, G. I., Proc. Roy. Soc. 145A (1934), 362.

28. Burgers, J. M., Proc. Kon. Nederl. Akad. Wetensch. 42 (1939), 293, 378.

29. Kröner, E., *Theory of Crystal Defects*, ed. by Boris Gruber, Proc. Summer School Hrazany 1964 (Czech. Acad. of Sciences, 1966).

30. Hirth, J. G., *Mathematical Theory of Dislocations*, ed. by Toshio Mura (ASME, 1969).

31. Kondo, K., Proc. 2nd Japan Congr. Appl. Mech., 1952, 41.

32. Bilby, B. A., R. Bullough and E. Smith, Proc. Roy. Soc. Lond., A231 (1955) 263.

33. Golebiewska-Lasota, A. A., Int. J. Engng. Sci. 17 (1979), 329.

34. Harris, W. F., South African J. Sci. 74 (1978).

35. Morse, P. M. and H. Feshbach, *Methods of Theoretical Physics* (McGraw-Hill, New York, 1953).

36. Kachanov, L. M., *Foundations of the Theory of Plasticity* (North Holland, Amsterdam, 1971).

37. Golebiewska-Lasota, A. A. and D. G. B. Edelen, Int. J. Engng. Sci. 17 (1979), 335.

38. Edelen, D. G. B., Int. J. Engng. Sci. 17 (1979), 441.

39. Edelen, D. G. B., Int. J. Solids Structures 17 (1981), 729.

40. I. M. Gelfand and S. V. Fomin, *Calculus of Variations* (Prentice-Hall, Englewood Cliffs, New Jersey, 1963), 165ff.

41. Edelen, D. G. B., Int. J. Engng. Sci. 14 (1976), 1013.

42. Edelen, D. G. B., Ann. Phys. (N.Y.) 112 (1978), 366.

43. Bataille, J. and J. Kestin, J. Non-Equilib. Thermodyn. 1 (1976), 25.

44. Edelen, D. G. B., J. Non-Equilib. Thermodyn. 2 (1977), 205.

45. Bataille, J., D. G. B. Edelen and J. Kestin, Int. J. Engng. Sci. 17 (1979), 563.

Lecture Notes in Physics

Vol. 136: The Role of Coherent Structures in Modelling Turbulence and Mixing. Proceedings 1980. Edited by J. Jimenez. XIII, 393 pages. 1981.

Vol. 137: From Collective States to Quarks in Nuclei. Edited by H. Arenhövel and A. M. Saruis. VII, 414 pages. 1981.

Vol. 138: The Many-Body Problem. Proceedings 1980. Edited by R. Guardiola and J. Ros. V, 374 pages. 1981.

Vol. 139: H. D. Doebner, Differential Geometric Methods in Mathematical Physics. Proceedings 1981. VII, 329 pages. 1981.

Vol. 140: P. Kramer, M. Saraceno, Geometry of the Time-Dependent Variational Principle in Quantum Mechanics. IV, 98 pages. 1981.

Vol. 141: Seventh International Conference on Numerical Methods in Fluid Dynamics. Proceedings. Edited by W. C. Reynolds and R. W. MacCormack. VIII, 485 pages. 1981.

Vol. 142: Recent Progress in Many-Body Theories. Proceedings. Edited by J. G. Zabolitzky, M. de Llano, M. Fortes and J. W. Clark. VIII, 479 pages. 1981.

Vol. 143: Present Status and Aims of Quantum Electrodynamics. Proceedings, 1980. Edited by G. Gräff, E. Klempt and G. Werth. VI, 302 pages. 1981.

Vol. 144: Topics in Nuclear Physics I. A Comprehensive Review of Recent Developments. Edited by T.T.S. Kuo and S.S.M. Wong. XX, 567 pages. 1981.

Vol. 145: Topics in Nuclear Physics II. A Comprehensive Review of Recent Developments. Proceedings 1980/81. Edited by T. T. S. Kuo and S. S. M. Wong. VIII, 571-1.082 pages. 1981.

Vol. 146: B. J. West, On the Simpler Aspects of Nonlinear Fluctuating. Deep Gravity Waves. VI, 341 pages. 1981.

Vol. 147: J. Messer, Temperature Dependent Thomas-Fermi Theory. IX, 131 pages. 1981.

Vol. 148: Advances in Fluid Mechanics. Proceedings, 1980. Edited by E. Krause. VII, 361 pages. 1981.

Vol. 149: Disordered Systems and Localization. Proceedings, 1981. Edited by C. Castellani, C. Castro, and L. Peliti. XII, 308 pages. 1981.

Vol. 150: N. Straumann, Allgemeine Relativitätstheorie und relativistische Astrophysik. VII, 418 Seiten. 1981.

Vol. 151: Integrable Quantum Field Theory. Proceedings, 1981. Edited by J. Hietarinta and C. Montonen. V, 251 pages. 1982.

Vol. 152: Physics of Narrow Gap Semiconductors. Proceedings, 1981. Edited by E. Gornik, H. Heinrich and L. Palmetshofer. XIII, 485 pages. 1982.

Vol. 153: Mathematical Problems in Theoretical Physics. Proceedings, 1981. Edited by R. Schrader, R. Seiler, and D.A. Uhlenbrock. XII, 429 pages. 1982.

Vol. 154: Macroscopic Properties of Disordered Media. Proceedings, 1981. Edited by R. Burridge, S. Childress, and G. Papanicolaou. VII, 307 pages. 1982.

Vol. 155: Quantum Optics. Proceedings, 1981. Edited by C.A. Engelbrecht. VIII, 329 pages. 1982.

Vol. 156: Resonances in Heavy Ion Reactions. Proceedings, 1981. Edited by K.A. Eberhard. XII, 448 pages. 1982.

Vol. 157: P. Niyogi, Integral Equation Method in Transonic Flow. XI, 189 pages. 1982.

Vol. 158: Dynamics of Nuclear Fission and Related Collective Phenomena. Proceedings, 1981. Edited by P. David, T. Mayer-Kuckuk, and A. van der Woude. X, 462 pages. 1982.

Vol. 159: E. Seiler, Gauge Theories as a Problem of Constructive Quantum Field Theory and Statistical Mechanics. V, 192 pages. 1982.

Vol. 160: Unified Theories of Elementary Particles. Critical Assessment and Prospects. Proceedings, 1981. Edited by P. Breitenlohner and H.P. Dürr. VI, 217 pages. 1982.

Vol. 161: Interacting Bosons in Nuclei. Proceedings, 1981. Edited by J.S. Dehesa, J.M.G. Gomez, and J. Ros. V, 209 pages. 1982.

Vol. 162: Relativistic Action at a Distance: Classical and Quantum Aspects. Proceedings, 1981. Edited by J. Llosa. X, 263 pages. 1982.

Vol. 163: J. S. Darrozes, C. Francois, Mécanique des Fluides Incompressibles. XIX, 459 pages. 1982.

Vol. 164: Stability of Thermodynamic Systems. Proceedings, 1981. Edited by J. Casas-Vázquez and G. Lebon. VII, 321 pages. 1982.

Vol. 165: N. Mukunda, H. van Dam, L.C. Biedenharn, Relativistic Models of Extended Hadrons Obeying a Mass-Spin Trajectory Constraint. Edited by A. Böhm and J.D. Dollard. VI, 163 pages. 1982.

Vol. 166: Computer Simulation of Solids. Edited by C.R.A. Catlow and W.C. Mackrodt. XII, 320 pages. 1982.

Vol. 167: G. Fieck, Symmetry of Polycentric Systems. VI, 137 pages, 1982.

Vol. 168: Heavy-Ion Collisions. Proceedings, 1982. Edited by G. Madurga and M. Lozano. VI, 429 pages. 1982.

Vol. 169: K. Sundermeyer, Constrained Dynamics. IV, 318 pages. 1982.

Vol. 170: Eighth International Conference on Numerical Methods in Fluid Dynamics. Proceedings, 1982. Edited by E. Krause. X, 569 pages. 1982.

Vol. 171: Time-Dependent Hartree-Fock and Beyond. Proceedings, 1982. Edited by K. Goeke and P.-G. Reinhard. VIII, 426 pages. 1982.

Vol. 172: Ionic Liquids, Molten Salts and Polyelectrolytes. Proceedings, 1982. Edited by K.-H. Bennemann, F. Brouers, and D. Quitmann. VII, 253 pages. 1982.

Vol. 173: Stochastic Processes in Quantum Theory and Statistical Physics. Proceedings, 1981. Edited by S. Albeverio, Ph. Combe, and M. Sirugue-Collin. VIII, 337 pages. 1982.

Vol. 174: A. Kadić, D.G.B. Edelen, A Gauge Theory of Dislocations and Disclinations. VII, 290 pages. 1983.